A RANCHMAN'S RECOLLECTIONS

The Fred H. and Ella Mae Moore Texas History
Reprint Series

The Texas Revolution by William C. Binkley (1979)

Notes Taken During the Expedition Commanded by Capt. R. B. Marcy, U.S.A., Through Unexplored Texas, in the Summer and Fall of 1854, by W. B. Parker (1984)

Spanish Explorers in the Southern United States, 1528–1543, edited by Frederick W. Hodge and Theodore H. Lewis (1984)

Texas, by Mary Austin Holley (1985)

A Ranchman's Recollections by Frank S. Hastings (1985)

The publication of this series is made possible by a gift from Fred H. and Ella Mae Moore and the Mobil Foundation, Inc.

A RANCHMAN'S RECOLLECTIONS

An Autobiography

IN WHICH UNFAMILIAR FACTS BEARING UPON THE ORIGIN OF THE CATTLE INDUSTRY IN THE SOUTHWEST AND OF THE AMERICAN PACKING BUSINESS ARE STATED, *and* CHARACTERISTIC INCIDENTS RECORDED

BY

FRANK S. HASTINGS

Manager of the S. M. S. Ranch

STAMFORD, TEXAS

Published by

THE TEXAS STATE HISTORICAL ASSOCIATION

In Cooperation with the

CENTER FOR STUDIES IN TEXAS HISTORY

THE UNIVERSITY OF TEXAS AT AUSTIN

Library of Congress Catalog Card Number: 85-051948

ISBN 0-87611-077-4 Hardback Edition
0-87611-078-2 Paperback Edition
0-87611-079-0 Special Limited Edition

The past is open to us. It is our key to understanding. If we look to our history we can see how our ancestors cut the brush for us and made our way in the world possible. We can see how we arrived where we find ourselves today. We can see that life is a process of change, and that we must know our past to understand the change.

Fred H. Moore

The Barker Texas History Center, The University of Texas at Austin, provided the copy of *A Ranchman's Recollections* used to print this edition.

INTRODUCTION

The tape recorder was turning, and Clifford B. Jones carefully chose his words as he assessed the men he knew. A long career in West Texas farm and ranch management, banking, and education had made him one of the most knowledgeable people on the region's colorful past and energetic present. It was the spring of 1968 when Texas Tech archivist David B. Gracy asked Jones about his old friend.

"Frank Hastings was one of the greatest ranchmen I have ever known," Jones said emphatically. It was a statement easily made, for Jones had observed firsthand the last fourteen of Hastings's twenty successful years as manager of the Swenson ranching interests in West Texas. At the time, the Swenson operation, known popularly as the SMS, was one of Texas's largest ranches, and, thanks to Hastings, one of the most well known in the nation.

The SMS heritage began with one of the state's most colorful pioneers. Swen Magnus Swenson, perhaps Texas's first Swedish immigrant, made a sizeable fortune as an Austin merchant prior to the Civil War. He invested in railroad and land certificates which gave him title to large tracts of land in then unsettled West Texas. As the "beef bonanza" swept through the Great Plains in the late 1870s and early 1880s, Swenson decided to put his vast holdings to work by leasing his land to his two sons, Eric Pierson and Swen Alvin, who in turn established the SMS ranches. The Swenson brothers purchased additional land and eventually increased their holdings to include four 100,000-acre ranches: the Throckmorton, located in Throckmorton County; the Flat Top, in Jones, Stonewall, and Haskell counties; the Tongue River, in Motley and Cottle counties; and the Spur, in Garza and Crosby counties.

Like many of the vast Texas ranches, the SMS was not profitable in its early years. Cattle bought at record high prices in 1882 could not produce profitable calves when the market collapsed in 1886. The dry years of 1886 and 1892–1894, coupled with the national depression of 1893, kept cattle prices low and forced many ranches out of business. The Swensons, however, encouraged by rising land values, were determined to succeed and continued to buy land.

Also, by the turn of the century, Americans were demanding higher quality beef than the rangy grass-fed Texas longhorn. Midwestern stock feeders, utilizing such improved breeds as the Hereford and Angus, had captured the market by fattening cattle on corn in their feedlots. The Swenson brothers, hoping to improve their own market, sought a manager for their ranches who was an experienced breeder and feeder. They found their man in Frank Hastings.

A native of Leavenworth, Kansas, Hastings had attended Notre Dame and the University of Michigan. In 1889 he had begun a thirteen-year career with the famous Armour family, working first as a supervisor of meat brokers in the northwestern United States. Then in 1892, at the request of Kirkland B. Armour, he joined the Armour purebred Hereford operation at Excelsior Springs, Missouri, near Kansas City. Working with the advice of the top Hereford breeders in the nation, Hastings soon became responsible for the herd's breeding program and sales. In 1902, when Hastings moved to Texas from Missouri, he was already widely respected by midwestern and western cattle interests.

Hastings was forty-two when he became the Swensons' manager. He brought to the SMS headquarters at Stamford a cosmopolitan "gentleman farmer" attitude which widely influenced his community. And he introduced Texas ranchers to two major innovations. First, he pioneered the development of large-scale cattle feeding in West Texas by constructing a 5,000-head feeding-pen complex at Stamford, where cattle were fattened on cottonseed meal and molasses. Second, he developed the concept of a "mail order" sale business by persuading midwestern stock feeders to order SMS cattle sight unseen. To promote the Swenson product, Hastings produced attractive, well-illustrated brochures which depicted ownership, management, cattle, range, and daily life on the SMS ranches. He later expanded his audience by writing articles for regional and national livestock newspapers. This book, *A Ranchman's Recollections*, was originally written as a series of articles for the *Breeder's Gazette* in 1920.

The reader should be advised that Hastings's book is not the typical romantic glorification of life in the saddle. It offers instead an inside glimpse of other aspects of the cattle business, and includes such topics as breeding, packing, and promoting cattle. But it does contain some wonderful insights into life on a ranch, and the last chapter, "The Speckled Yearlin'," is unforgettable.

During his twenty-year tenure as the Swenson manager, Hastings transformed the SMS operation into an innovative leader and champion cattle producer. In 1918 a carload of SMS cattle won Grand Champion of the Chicago International Stock Show, an event that marked the first time commercial cattle had won over purebred lots. The prize made the SMS famous. But Frank Hastings did not have time to exploit fully the ranch's success. He died in Stamford in June 1922, a week before his sixty-second birthday.

Although no imposing monument commemorates the life of Frank Hastings, this little book will continue to serve as a tribute to a great cattleman. Perhaps that is what Clifford Jones had in mind when, in 1947, he helped in the founding of the Southwest Collection at Texas Tech and donated as its first book Frank Hastings's *A Ranchman's Recollections*.

David J. Murrah
Texas Tech University

A RANCHMAN'S RECOLLECTIONS

This book published and for sale at the office of *The Breeder's Gazette*, 542 South Dearborn Street, Chicago, Ill.
Price $1.75 postpaid.

Sketch by Frank Tenney Johnson

"I guess we must a' bin three feet ahead, with a hundred yards to go. Split was a-cussin' an' spurrin', an' whippin'. I jes' whispered, 'Now, Stud! Now! Now!'"

(See Chapter XXIV—"Old Gran'pa")

A RANCHMAN'S RECOLLECTIONS

An Autobiography

IN WHICH UNFAMILIAR FACTS BEARING UPON THE ORIGIN OF THE CATTLE INDUSTRY IN THE SOUTHWEST AND OF THE AMERICAN PACKING BUSINESS ARE STATED, *and* CHARACTERISTIC INCIDENTS RECORDED

BY

FRANK S. HASTINGS

Manager of the S. M. S. Ranch

STAMFORD, TEXAS

PUBLISHED BY

THE BREEDER'S GAZETTE

542 SOUTH DEARBORN STREET

CHICAGO, ILLINOIS

1921

The Lakeside Press
R. R. DONNELLEY & SONS COMPANY
CHICAGO

TO THE OWNERS AND EMPLOYEES
OF THE S. M. S. RANCH
WHO THROUGH THEIR CO-OPERATION AND
UNSWERVING CONFIDENCE
HAVE GIVEN ME INSPIRATION IN MY
INDUSTRIAL WORK
THIS LITTLE VOLUME IS
AFFECTIONATELY DEDICATED.
FRANK S. HASTINGS

PUBLISHER'S PREFACE

"And thou shalt smite the rock, and there shall come water out of it, that the people may drink."

IN the opening paragraph of the first chapter of this volume the author intimates that we "may" not have known that we were to unloose a flood when we asked him originally to write of his experiences in connection with the rise and progress of the western cattle trade. In response I will simply say that we have known FRANK HASTINGS for full thirty years. We thought we knew him well, but we now find that he has gifts not hitherto suspected.

Originally brought out in the columns of *The Breeder's Gazette,* these sketches attracted at once and held the close attention of thousands of delighted readers in every part of the country. In soliciting their preparation we approached the subject with the same confidence that the Prophet of old smote with his rod the rock of Horeb; but I must here and now confess that the stream of thought called forth surpassed in purity and sweetness anything anticipated.

Mr. HASTINGS has made a real contribution to the pastoral literature of the West; one vitalized throughout by an intensely human touch. We count ourselves fortunate in having the privilege of giving his work this permanent form.

ALVIN H. SANDERS.

AUTHOR'S PREFACE

THE only excuse I can offer for the publication in book form of the series of sketches which under the title, "Recollections of a Ranchman," appeared in *The Breeder's Gazette,* Chicago, from July 15, 1920, to and including the Holiday Number of that year, is found in letters from people whom I know and many whom I do not know, and from librarians and educational institutions, suggesting that the matter should be republished in this form.

I confess that while the opinions of my immediate friends are dear to me, yet they may be indulgent and that the kind words written from the forks of the creek by people whom I do not know have been a greater boon to me. This is particularly true of old-timers on the range who have commented on the cowboy dialect used in some of the sketches as being correct, or, as one or two have written, "You have got the cowboy down fine."

It is solace to a writer to have his written word ring true to "men who know." Their expressions of appreciation have gone straight to my heart.

It has been a source of pleasure to look backward over the cattle industry as it has come into my life, and to recall some of the wonderful men whom it has been my privilege to know. This prelude would not be complete without an expression of my gratitude to Alvin H. Sanders and DeWitt C. Wing of *The Breeder's Gazette,* for their patience, care and thoroughness in editing my manuscript, and their helpful suggestions as to topics.

If by chance an indulgent reader shall find anything of historic value in these sketches, or if the book shall prove helpful to any of the young men who are coming along in the cattle industry, my cup will be full.

Very truly,

FRANK S. HASTINGS,

Manager of the S. M. S. Ranch,

Stamford, Texas.

June 1, 1921.

CONTENTS

LIST OF ILLUSTRATIONS

CHAPTER I

THE EARLY DAYS

IT IS a far cry from the S. M. S. Ranch in 1920 to the five-board fence on my father's little farm, near Leavenworth, Kans., where I sat fifty years ago (in 1870) watching the first herd of Texas cattle that I had ever seen. They were driven into an adjoining pasture, and as I look back through the vista of men, methods and events that have filled that gap in the evolution of the cattle industry I am reminded that *The Breeder's Gazette* in asking me to review that period from a personal standpoint may not have realized how full of it I am, or how much space would be required to tell the story, which must necessarily be rambling. I shall not attempt to write history in chronological order. As Henry Watterson, in some wonderful stories of men and events, in his "Looking Backward" series, wrote of characters and events of interest rather than orderly history, so I wish I might use the general title of his recollections, because my own will be simply the memories of "looking backward," and yet I shall try to reach the time when the past steps upon the threshold of the present.

Harking back to the boy of 10 years on the board fence, I recall that the cattle were from the

King Ranch, and had been shipped from Abilene, Kans., which was then the great objective of Texas trail herds, and the first great shipping point in the west. It was late in the fall of 1870. The herd was of mixed cows and steers. The steers in the main were disposed of by December and the cows wintered. Their pasture had been my playground. The "Ole Swimming Hole" was there; it was also my wild duck preserve, in fall and spring, as well as my fish pond. Black walnuts lined the stream and hazelnuts abounded on the rolling ground; they, too, were my treasures. My winter rabbit trapping and quail shooting were there. I used the scrub oak thickets for a screen from which to shoot prairie chickens as they came over in the fall on their way southward.

The longhorns seemed an intrusion. They had a wild-eyed way of keeping one covered. I was inclined to resent their advent, but my mother said that they would not bother me if I attended to my own business and let them alone. Perhaps that was intended as much for a life admonition as to remove my misgivings. It worked out all right, and I was soon *persona grata* to go and come. I have often thought of her admonition, with its practical demonstration; it has helped me over some hard spots since.

In the following spring the Texas cows began to bring calves. The town butchers wanted them, and the cattle-owners wanted the cows dry and grass-fat. The calves were sold at $2.50 per head, the butchers to come and get them, as wanted. Their first trip, made in an ordinary butcher's wagon, with no horse to rope from, encountered maternity from the jump.

My father had a first-model Smith & Wesson blue-barreled six-shooter. He had furnished me with cartridges for unlimited practice. I have not carried or fired a revolver in forty years, but at that time I could pepper my shots pretty well about the ace at fifty feet. I hired out to the butchers, and had good success shooting the calves in the forehead from the wagon. I was allowed 10 cents per head, which was probably a better return, in proportion, than the cattle-owners, received from their investments.

In the summer tragedy came. The winter had been very mild; ticks carried over; a dairy herd which had been culled and selected for years was turned into the pasture with the Texas cows, and suffered a 90 per cent mortality. The whole country became excited, thinking it an epidemic which would spread; stories of Texas fever were brought in vaguely from other states; a great many domestic cattle in Dickinson County, Kans., near Abilene, died; New York State issued a quarantine against Texas cattle; the Governor of Illinois called a convention at Springfield, Ill., which was attended by delegates from most of the northern states and two from Canada; Texas fever and quarantines were discussed. Only three theories, with any following, were presented. The first was a theory advanced by scientists which argued that Texas fever resulted from a small egg or sporule, deposited upon the blades of grass in Texas. The blades, being eaten by the cattle, enabled the sporule to find its way into the blood and grow to be well-defined, under the microscope, resulting in disorganization of the blood and a fever of deadly character. The second theory was that Spanish or Texas fever came from such causes as fever aboard emigrant ships, privation, hard usage, and insufficient feed, water and rest. The third theory was ticks, which, dropped from Texas animals, were eaten by domestic animals, with fatal results.

Texas was without representation at the meeting, from which prejudice was spread with more or less justification over the whole United States against Texas cattle, setting the industry back many years. About this time the nucleus of the now great packing industry began to form, but for the moment I shall only write of it as it came into my own life.

One Joseph Whittaker of Cincinnati gave a 99-year lease on some valuable vacant property for city park purposes. He received a yearly income which looked big then, and, with cattle and hogs at low value, furnished a good working capital, protected by a handsome annual income. He came to Leavenworth, Kans., formed a partnership with Matthew Ryan, and built a substantial small packinghouse for the slaughter of hogs and cattle on the bank of the Missouri River, which stream formed a cheap sewer as well as cheap ice supply. They specialized on hogs during the winter months, and cattle during the cattle season; that is, from mid-summer to December, doing a limited fresh meat business, and running largely to barreled beef and dried beef hams.

A lot of negro women and children could always be seen coming from the packinghouse, carrying fresh livers, hearts and kidneys, which were given away. In modern technical parlance these organs were the "pluck," which is now a material item in modern packinghouse salvage. In looking back I do not understand why it was not thrown into tankage, which they were then making, nor why they did not use in the main everything else in the offal.

Whittaker bought a half-section of land near my father's place. It was a rather light-soiled tract. He began hauling tankage to scatter over it. That was my first sense of the packer drive to eliminate waste. It was literally a sense, because the first intimation came through the nostrils and the second through the ears from the expressions of outraged neighbors, who sought the courts and tried to have it declared a nuisance, but most of them lived to see an almost sterile farm develop into one of the best producers in the country. I do not know whether they ran the blood into tankage then or not, but have a vague memory that it went into the sewer. What has followed in the wake of the primitive initial move in animal fertilizer production and use, in comparison with what is being accomplished now, and the greater saving from using what originally went into tankage, will come in for later comment.

Whittaker's boys were my playmates and chums. On Saturdays we often went to the packinghouse. On one of these visits an incident occurred which has been vital in relation to some of my subsequent studies in breeding. They were killing quite a string of aged Texas steers, using a sharp lance, and striking behind the horns. We saw them lance a big fellow, with the usual result: a quick fall, the trap door opened, and he was dragged to the skinning beds. When the knife was at his throat he jumped, with one bound, to his feet, saw daylight through a door at the rear, jumped a story and a half to the ground, swam the Missouri River to a sandbar one-quarter of a mile distant, shook himself and turned his head to the shore, at bay. In later years Texas steers running amuck in the Kansas City bottoms charged cable cars, head on.

I am sure that the wonderful vitality of the primitive Longhorn, backed by self-reliance and the hustling qualities of the Hereford, blended with the old Spanish blood, has served as a kind of iron basis for the well-bred bovine stocks in Texas today. This vital seed, as represented by its modern beef type, transplanted to the north, stands the rigors of winter better than native cattle. In my own work here in Texas I found that the S. M. S. herd had a wonderful basis. There were many weeds in it, however, and I cut for conformation, leaving lots of 1,000-pound cows, with some brindle or often straight duns, long after I had cut out a lot of cat-hammed, flat-ribbed animals with clean flesh marks and color.

Cortez may have treated the Mexican race badly, but he gave to America or, more exactly speaking, to the Texas prairies, a heritage in vital cow brutes which has done almost as much as pure breeding for the American cattle industry. That heritage furnished the vitality in which to fix the beef-making instinct. I have never lost sight of it in the S. M. S. herd. While our fraction long ago was reduced to .999 pure, we have kept our eyes on the strong, "good-doers," and watched the winter feed grounds, like hawks watching chickens, for weak constitutions, and the spring pick-up for the laggards, and kissed them goodbye.

We owe the primitive Texas cow a debt. She had much to do in making possible the record that well-bred Texas cattle are registering in the markets, as regards both prices and poundage.

If the foregoing may take the caption "Fifty Years Ago," perhaps what follows could be headed "Light Begins to Dawn."

The early days of range cattle production may only be considered as a tragedy, in the light of what has followed—a thought which can probably be applied with truth to the frontier in making a path for civilization and progress—but the story, rich in terrors and privations, of the gold-seeker and the trail from the Missouri River to the Golden Gate, has a sequel in the story of the men who have made the Texas of today.

From Charles Jones, an associate in my packing-house days, and now at the head of the Freeport, Tex., Sulphur Co., I have obtained a book which induced him to become a cowpuncher in 1877. It is out of print and priceless, and I go to the safe to see if it is still there, just as one puts one's hand on one's pocketbook in a crowd. It was written by "Joe" G. McCoy, and bears the title "Historic Sketches of the Cattle Trade of the West and Southwest." It was published in 1874, just at a time when the dark hour was on the Texas industry. The state was under quarantine, and there was no known remedy or protection against Spanish or Texas fever. The book contains interesting data. Its most interesting chapters deal with attempts to find an eastern market for cattle, first by trail, even into Illinois, and in 1868 to Abilene, Kans., when shipping pens had been established on the Union Pacific Railroad and buyers assembled during the season for purchases distributed by rail everywhere east.

Conditions were at their worst in 1873, when Mr. McCoy evidently turned in his copy. The wail of despair in his concluding thought sums up the history of those awful preceding ten years. I quote him, not only because of what has followed, but because the evolution of Texas into a nursery for well-bred beef cattle in the national supply offers a striking illustration of the courage and never-know-when-you-are-licked-persistence of the Texan, in the cattle industry or any other industry. Texas is the first state in cotton and cattle; the third in oil, with the prospect of becoming first; the first in sulphur, and wheat is coming fast, to say nothing of truck gardening, poultry and butter—all distinct national factors—and there are still millions of fertile acres of virgin soil begging for the plow; but I am getting away from my story. Mr. McCoy, himself a pioneer frontier trailer and cattle trader, says, in 1874:

"Of the cattle coming from Texas two-thirds are marketed when almost totally unfit for consumption, thus entailing comparatively immense losses upon the parties selling them. Rather than continue this foolish, wasteful and ruinous practice, drovers had infinitely better buy stock ranches in western Kansas and Colorado and there keep them until their stock is fat."

It is not my thought to ridicule McCoy's conclusion. He was logical and could not look forward to the Texas of today, any more than the cattlemen of those times could look forward to agricultural development and solidify their land holdings when the alternate section was school land under lease and it seemed cheaper to lease than to buy.

The nester was not dreamed of then. The range was open, and the wire fence was still in the back of John Gates' head. There were a few notable exceptions. S. M. Swenson was one of them, and we find on his maps this notation, made before the Civil War: "Do not sell this tract; it has water on it, and is good agricultural land." Men who had that foresight have reaped deserved rewards.

The northwestern movements had not begun to any appreciable extent, but except in so far as the Spanish fever scare limited trails to Kansas railroad connections, a ray of light had appeared at Kansas City, and there was another day dawning for the beef industry. That light was the packinghouse plus refrigeration. First came the packinghouse, with its immense capacity for barreled beef and pickled dried beef hams, with a limited natural ice refrigeration. Barreled beef was almost as great a staple as barreled pork, and it had an immense consumption in lumber camps and sailing vessels, besides a large aggregate domestic consumption.

Plankington and Armour owned packinghouses in Milwaukee and Chicago, but with that great foresight which seemed to make the future an open book, and made a fortune for him in pork after the

Civil War, Philip D. Armour became convinced that he needed a location nearer the source of beef production. Accordingly in 1870 he located a packinghouse in Kansas City, devoted largely to the slaughter of beef, although the hog was an important factor in the business. The Kansas City stockyards were simply a holding-place. Crude cuts of the yards in the McCoy book show them to be made like good-sized railroad feed and rest pens of the present day, probably about twice the size of those at Parsons, Kans.

Some trading was done in the yards, but in the main packinghouse buyers went out to Abilene, taking real money with them, and making their purchases as trail herds came in. G. W. Tourtelotte, familiarly known as "Charlie" Tourtelotte, later superintendent of the Kansas City plant for twenty years, was Armour's pioneer buyer. I wish that space permitted my going into details concerning the reminiscences that I have heard him relate. He was a man of unusual balance, likable, competent and dependable. He did not believe, even in those early days, when it was so much in vogue, in mixing "booze" and business. From others I have learned that his great asset with the men whom he came in contact with in those early days was his absolute fairness. He established for his company that most valuable of all business assets: "A good concern to do business with."

"Charlie" was a born handler of men, always taking the human equation into consideration, and, during his superintendency, probably had as few labor complications as any superintendent in any industry in America. His men idolized him. During a strike, which came very quickly after the plant was visited by labor organizers, I recall that men were hard to hold in line, and persisted in telling the agitators that they would do anything "Charlie" Tourtelotte said was right. I recall that Samuel Gompers came promptly and effected a settlement. The strike was with the firemen, and lasted only a short time; but with its inception notices were sent out all over the country urging a boycott of the products of the Armour Packing Co., which had not then consolidated with Armour & Co. of Chicago, with Philip D. Armour as a large stockholder.

Corrective notices were of course sent. At that time the western mining districts were strongly organized. Butte, Mont., the most notable, was the leading district, and at that place there were 52 organizations, operating through a central council, and a notice of the lifting of the boycott had not reached that body. Everything was peaceful at Butte. In the Coeur d'Alene country, with headquarters at Wallace, Idaho, practically the same condition as to organization existed; but in that country things were very bad. Mine-owners were being killed, and troops were sent there.

At this point I shall digress for a moment to illustrate the workings of the boycott system in restraint of trade, and how far-reaching, after local troubles had been adjusted, the system sometimes unintentionally carried serious injury, and relate some personal reminiscences which are amusing in the telling, but were very real then.

I had charge of the western brokers, and usually spent two months every spring in the west. Our immense trade there was blocked by the failure to have the boycott raised. Our brokers were wiring us about it. I was sent out to investigate. I found that the boycott had not been raised because the notice of its having been lifted had not been placed before the central body, the secretary of which told me that it would be necessary to locate the particular union which presented it. His record did not show that union, so I had to make the rounds, of the 52 secretaries until I located my man. I ferreted out two or three of the leaders. They treated me very courteously, and helped me locate the various secretaries. Butte was a three-shift town, most of whose population consisted of miners; it was probably the liveliest town in America at that time. It had no day or night. Members of one shift or another were always more or less on the streets. It was said that "It is day all day in the daytime, and there is no night at Butte."

I began selling meats, subject to lifting the boycott, and spent a week, most of the 24 hours each day, as the shifts were out, trying to locate my man. Most of them dropped into Demer & Hicky's saloon, and I spent most of my time there playing whist with some of the good fellows who would point out some of my 52 varieties as they came in, but always the wrong man. Finally I got down to five, and located them in certain mines. I was doing a land office business selling carloads, but they were needing the goods and punching me up for shipment, which we did not dare make until my man was found.

The central body had its monthly meeting the next night. I went out to one of the mines where one of my five men worked. It really was not a mine but a hole in the ground about 500 feet deep, with a bucket instead of a cage in which to bring the ore up, and the bucket, by the way, was a whisky barrel, with a steel-pointed bottom. My man was there, but down below. The engineer was a fine fellow, but he evidently wanted to have some fun, so he said, "You can go down and see him." I asked if the man could not come up for a few minutes, and said that I would pay his wages for the day. The engineer said, "You are not afraid to go down, are you?" I replied, "No," but I lied, and could feel myself slipping off the barrel, yet I knew that I must have the miner's respect. He gave me some slicker clothes and delegated one of his helpers to accompany me.

We stood with our feet on the rim of the barrel, our hands on the cable, and as the drop began I looked up at the daylight above, and said, "Goodbye, old girl, if I never see you again."

Below the level of the mine there was a sump, and the engineer, not through with his fun, dropped us to the water line before my guide could signal him to stop. I found my man, got my release, and when I reached the top was so glad to see any one but the engineer that I rushed over and kissed a handsome setter dog. That night the central body met and released us, but the miners told the story, and had about as much fun over it as boys on a ranch have when some one is thrown but not hurt.

I shall not here enter into a description of very much the same procedure in the Coeur d'Alene country, except to say that I was accompanied by a merchant from Missoula, Mont., who had a branch at Wallace, Idaho. We went into a little restaurant in the mining district for lunch. There was a local strike on, and almost everyone had "tanked up." Our waiter had achieved a particularly comprehensive "jag." Some one pointed me out as the man who was trying to lift the Armour boycott. We ordered boiled eggs. The waiter called back, "Boil two dozen," and then addressing us directly, said, "When any of ye damned capitalists come out here to run this country we'll feed ye well, but, by God, ye'll eat what's set before ye."

My friend turned to me and said, "What are you going to do?"

"I am going to eat mine," I answered.

In the meantime I suggested to the waiter that we all go in at the next door and have a drink with the "capitalists." In the end the waiter found my man for me.

There is hardly a day in my life when something in connection with my old packinghouse training does not help me over a difficulty. A duty to perform meant in those days "sticktoitiveness," quick wits, keeping one's temper, attending to one's own business, and work. That is how men have been trained to encircle the globe with American live stock products.

CHAPTER II

EARLY PACKING AND REFRIGERATION

L. J. CALLAHAN in his reminiscences shows that Armour slaughtered the first beef, from a packing standpoint, in Kansas City in 1869, using the Nofsinger House, but in 1870 the company killed in its own plant, and by the fall of 1871 thought it remarkable to be killing 100 cattle in a ten-hour day. McCoy records 68,000 cattle killed in 1871 and 1872 in Kansas City, Armour killing one-half, the other one-half by others whom he does not mention; also that in 1873 26,800 were killed, the volume being reduced by the panic of that year. Callahan says that in 1869 some cattle were slaughtered in Houston, Tex., by Hancock & Cragin and some at Junction City, Kans., by Patterson & Co.

In 1869 Philip D. Armour sent P. C. Cole to Texas with a view to selecting a killing point, but after careful investigation Mr. Cole reported that it was too early to go to Texas, and, placing his finger on the map at Kansas City, said, "This is the logical point." Mr. Cole was more or less with the Armour Kansas City plant and was associated with Geo. W. Tourtelotte in the early range buying.

McCoy records that in 1873 the Missouri, Kansas

& Texas Railroad was completed to Denison, Tex., where pens for 2,000 cattle were built; also that the Atlantic and Texas Refrigerator Car Co. had constructed 100 new cars, adapted to shipping fresh beef and local capacity for killing 500 cattle per day. I have been unable to ascertain the fate of the Denison enterprise, but hope to before my story is finished. The reference to cars can hardly mean refrigerators, since all records seem to point to 1875 and 1876, with Chicago taking the initiative.

Callahan records 1876 as the year in which the first killing was done in Kansas City for local butcher trade. He also comments on beef hams as going in large quantity in tierces to William Windsor, Liverpool, England, contracted for in advance, and to Jacob Dold, Buffalo, N. Y., who later established a large plant in Kansas City. Procter & Gamble of Cincinnati bought the tallow, Buffalo, N. Y., took the hides, and a Massachusetts concern the horns. Callahan makes another comment which explains in a way why the packing business moved rapidly between 1870 and 1880. He says: "It was usually considered that if a packer either owned or could rent a packinghouse, and had money enough to accumulate a suitable stock of cooperage, salt and saltpeter, and had a fair line with a good bank, he was ready to run the business."

The first record of refrigeration, still using natural ice, is given by Joseph Nicholson of Chicago as having been built in 1875 by him, the idea being taken from a small domestic icebox, known as the Fischer box. A royalty of 6 cents per square foot was paid for the use of the idea. A year later Mr. Nicholson felt that he could improve the Fischer box, and built an entirely different one, using his own ideas with success. Another interesting fact recalled by Mr. Nicholson is that prior to the refrigerator car, fresh meat, mainly pork, was shipped in barrels, with a link of stovepipe filled with packed ice in the center and meat packed about it, but it was never sent further than Aurora or Galesburg in Illinois. Pork tenderloins were then considered a by-product, and sold for 6 cents a pound. The lowest temperature obtained from this class of refrigeration was 34°.

The Bankers' Magazine for January, 1919, carries a story of the house of Swift & Co., which says: "Mr. Swift in 1876 perfected a refrigerator car that would carry dressed beef to eastern markets in prime condition. He built his own refrigerator cars because the railroads refused to build them. The same is true of Armour & Co., and others with well-defined refrigerator car service in 1878, by which time the prejudice against refrigerated meats had been overcome by the splendid quality of prime beef in prime condition available for every table in cities."

The history of refrigerator cars is especially interesting at this time, when bills in Congress are seeking to take them away from the packers, and yet it is a fair deduction that without private initiative the beef industry must have dragged on for years without markets. In 1881 I was with a produce house in Denver. Hundreds of cars of apples and potatoes were shipped in the winter from the Missouri River, there being stoves in the cars. It is a fair deduction, too, that the development of meat refrigerator cars brought the same service for produce, fish and fruit, most of it under private initiative, years before it would have come from the slow process of railroad development.

It is not my thought to follow the development of the Kansas City market for cattle through any tedious process of dates and figures, but rather to sketch epochs of vital bearing.

The hog industry does not call for any special mention of its advent, except that refrigerator cars began the movement for the sale of fresh pork cuts, with limited natural ice refrigeration for curing, a gradual increase in summer killing, the time-honored term "winter-cured," as applied to hams and bacon, becoming obsolete as mechanical refrigeration became definite.

Artificial, or, more properly speaking, mechanical refrigeration, meaning the use of the ammonia process, was first used by Kingan & Co., Indianapolis, in 1885. Its use in Chicago and Kansas City occurred in 1886 or 1887, but it did not come into widespread use as to butter, eggs, cheese, poultry and fruit until after 1890. It will be seen how far-reaching the refrigerator car was in advance of mechanical refrigeration in establishing a market for both fresh beef and pork in at least all the medium-sized and large cities.

The years 1870 to 1880 marked the period of great change from the uncertain to the certain, from "salt junk" to fresh meat distribution, and gave the real swing not only to the western packing industry but to a conception of better cattle on the ranges.

The Armour family were farmers. The old family plow, dating back to pre-revolutionary times, is still to be seen at the Chicago stockyards. The generation of packers, consisting of five brothers, became great captains of industry. Philip D. Armour founded Armour & Co., and brought all of his brothers directly or indirectly into the business. In 1869 he sent the oldest brother, Simeon B. Armour, to take charge of the Kansas City packing venture, and later another brother, A. W. Armour, to head the banking firm of Armour Bros. in Kansas City, afterwards merged into the Midland National Bank. A. W. Armour never became a packer, but his sons Kirkland B. Armour and Chas. W. Armour went into the business as boys. A. W. Armour brought to the great Kansas City banking house the instincts of the trained country banker, with its cordial intimacies, human interests and intuitive estimate of men. He was in every way a lovable character, with the faculty of attracting young men and helping them in their business problems. Many of the stalwarts in today's Kansas City activities grow fondly reminiscent of him and his influence on their lives.

When I went with the Kansas City Armours in 1889 Simeon B. Armour was still the active head, with John Mansur as confidential adviser. Mr. Mansur had in the '70's been of the firm of Slavens & Mansur in Kansas City packing, but withdrew to go with the Armours. He remained with them until his death. Hides and fertilizer were his immediate specialties; he also directed the hog buying. He was known throughout the industry as one of the ablest men of his time.

Among the traditions of the Kansas City plant is that of Simeon B. Armour's methods. He sat at the receiving scales in the early days, weighing in the hogs. He kept in close touch with the class of cattle and every day saw every nook and corner of the plant, keeping in the closest touch with his men and calling most of them by name, all distinct Armour traits. He was reserved, conservative and eminently just; his charities were extensive but never ostentatious. Himself a man of exemplary habits, he had the deepest sympathy for human weaknesses, and the cry of a soul in distress always found his hand reaching down to grasp one struggling in despair. He did not have the magnetic force of his brother Philip or of his nephew Kirkland, but by a quiet force, peculiarly his own, drew men to him with bonds equally strong.

When Kansas City, the unsightly town of hills and hollows, began to develop its parks and boulevards, he was on the first board, remaining until his death, and was the great inspiration for the now beautiful Kansas City. Mr. Armour attended church regularly, and had a quiet but effective method of passing on the sermon. If it appealed to him he remained and gave the minister a cigar, but otherwise filed out with the congregation. An instance of his innate honesty occurred when the celebrated heifer Armour Rose was being raffled off to build Convention Hall. Kirk B. Armour was making a public sale of registered Herefords, and several heifers sold at $1,000. Members of the Commercial Club asked him to donate a heifer to be sold for the Convention Hall fund, which he did, saying that she was as good as he had. The public knew the sale price, and figured that Armour Rose should be worth as much as the best, but very few wanted a heifer.

S. B. Armour called me to his desk and asked, "What is this that I see in the papers about Kirk and you calling this heifer worth $1,000? Did you say that she is worth that?" I said that she had been given without valuation, but the public had jumped at the conclusion that it could have the heifer or $1,000. He replied, "We'll make her worth a thousand dollars."

He had a weakness for hickory wood from the earliest days until he died; he just could not see an outstanding load of straight hickory wood come into the smokehouses without saying, "Send that up to my house." He always had a four or five years' supply ahead, and, so far as possible, heated his house with it.

Another incident illustrated his penchant for the quick liquidation of obligations. While at Hot Springs with his family Mr. Armour met Hiram Berry, the manufacturer of "Old Crow" whiskey. Shortly after Mr. Armour's return Mr. Berry sent him a case of "Old Crow," with the following doggerel, signed by Mr. Berry: "To Mr. S. B. Armour:

"Herein, Mr. Armour,
I send you a 'Charmer';
A real assuager of grief;
It is good for the 'inner,'
And when taken at dinner
Will go mighty well with your beef.

"Perhaps you don't know
That our real 'Old Crow,'
Is made on the perfectest plan;
It will cheer, it will cure,
And we warrant it pure,
As the meat you put up in a can."

I had introduced verse into our advertising, with rather good results, and was known as "The Rhymester." Mr. Armour brought the Berry effusion over to my desk and said, "I want you to send Mr. Berry a case of the best things we prepare. I want it to be the best case that ever went out of this shop. When a man thinks enough of me to fix me up like this I want to do something for him; then see what you can do about some poetry."

I manufactured the following, and between the two doses of rhyme they had good times showing it to their friends who were sharing the bounty. (I wonder if even millionaires would not be appalled at giving or receiving a case of "Old Crow" now!)

"To Mr. Hiram Berry:

"All Hail! Mr. Berry,
I am feeling quite cheery;
Have just swallowed a drop of 'Old Crow.'
If I'd a shadow of grief
'Tis my earnest belief
You would banish the presence of woe.

"This parcel of stuff
I trust is enough
To keep the gray wolf from your door.
If it isn't, please know
That one croak from 'Old Crow'
Will bring you a hundred-fold more."

This was signed "S. B. Armour." Perhaps it may seem trivial to introduce these little incidents in the lives of the great captains of industry, but in my study of men give me those who are intensely human, and try to get a laugh every day.

CHAPTER III

"THE STORY OF THE HEREFORDS"

I BEGAN with the Kansas City Armours in 1889, and remained with them until 1902. My advent in the cattle industry proper came about in an accidental way. I was on the provisions side of the packing business, holding down a desk of specialties, "mild-cured, selected hams and bacon," for our whole territory, and also in charge of western brokers, which latter work brought me in contact, largely by observation, with the northwestern ranges, although I met many of the owners. In 1892 registered Herefords were "in the dumps." There was no movement; breeders were in despair. The range was taking some bulls but at values below the cost of production.

Chas. M. Culbertson, Newman, Ill., a retired Chicago packer, had become interested in Herefords in 1877, accumulating, through importations and topping American sources, one of the greatest herds in history. Having decided to clean up, he visited his old friend Philip D. Armour, explained his plans, and asked Mr. Armour to instruct his cattle buyers to mark the registered cattle up for all they would stand for slaughter. Mr. Armour told him that it was a shame to kill that class of stock, but that he would weigh it up for all it was worth on the market, and Mr. Culbertson could add whatever per head price he thought fair, and transfer the pedigrees to Kirkland B. Armour, who owned a beautiful section-farm at Excelsior Springs, Mo. This was done, and Mr. Armour wired his nephew that the cattle were being shipped to him. Here I must pause to speak of that remarkable breeding basis.

In Alvin H. Sanders' book, *The Story of the Herefords,* on page 424, the imported cow Prettyface 5735 is described. By original Anxiety and out of a Longhorn cow, she was the wonder of England as a two-year-old in 1881, repeating her victories in 1882 over the American circuit. This cow came with the Culbertson purchase. Mr. Sanders comments: "Unlike many cows with distinguished show records, she made a wonderful record as a breeder, giving birth to eleven calves in ten years, none of them twins." Mr. Sanders may have included her calves after the time she came to Kirk Armour, but I do not think so. She gave birth to five or six calves after reaching Mr. Armour, among them Lady Prettyface and Lord Prettyface. The cow Prettyface offers a striking comparison between the show winners of 1882 and those of the past twenty years. She was 13 years old when she came to Mr. Armour, but lived, as I recall it, until 1900, prolific to the last. She was distinctly long-horned, very short-coupled and low-down; not a very large cow. Her horns today would probably decide a breeder not to show her, and that takes me to some range observations, leading up, among other things, to our decision to dehorn the S. M. S. herd in the face of all sorts of range traditions against it.

I found that cornbelt buyers, as they looked over the breeding herd, inclined to think that any animal carrying much horn development got it from primitive Texas longhorn, and when I argued that the Hereford was basically a breed with much horn development, as evidenced by the Armour importations, and that English breeders had been slow to follow the American plan of breeding to the modern droop horn, they looked askance. I have been a Hereford man always, but while I do not think Hereford horn development in any way a detriment, it has given the range producer of well-bred whitefaces lots of headache to hear or read the oft-repeated theory that the animal still carries too much primitive blood.

The Culbertson herd included many daughters of The Grove 3d; also the cow Marcie (by Waxwork) which produced 17 calves, including several sets of twins; Wiltona Grove by Lord Wilton—but I will not burden this story with references to all the famous bloodlines in that wonderful basis of the original Armour herd. The herd bull was Kansas Lad 36832 by Beau Real by Anxiety 4th and out of Bertha by Torro. Then, as now, the blood of

Anxiety 4th was eagerly sought. I recall that once, in talking with John Steward, I said: "John, what is the best line of blood to breed Anxiety on?" He replied, "Anything." He was right. I recall that in the first Armour sale he bought the bull Tempter with two sires and one dam strong in Anxiety and a The Grove 3d outcross. He used this bull for one season, and sold him at a then big price, but when the calves began to come he went back and rebought him. Steward was to my notion the best breeder of his time, and a wizard at "nicking." I am sure that his early death only prevented him from becoming America's greatest breeder.

I began to tell of the accident which drifted me into the cattle industry, but it is hazardous to drift a little without drifting a lot. The ten years of my association with registered Herefords are so full of reminiscences of men and cattle that I cannot hope to hold the patience of my readers longer than it takes to do some skimming.

Mr. Sanders' Hereford history is an orderly chronicle, and those who do not have it in their library should waste no time getting copies. I shall offer a few sidelights in which I had the personal touch, and for that reason shall not refer to many of the most noted factors in the evolution of the American cattle industry.

When Kirk Armour took on Herefords they were sent to the Excelsior Springs, Mo., farm, now owned by E. F. Swinney of the First National Bank of Kansas City. The farm was managed by Charles Wirt, one of the best all-around farmers and stockmen whom I have known. He was an unusually good feeder. Aged steers from the stockyards were sent over every year, and mares from the packing-house stables were sent over, bred, and returned for work until within safe foaling time. Wirt was not a pedigree man, except that his breeding records were infallible, but he was backward in his book-keeping. Mr. Armour went to the farm frequently, taking his two boys Watson and Lawrence, now vital factors in the Chicago plant. He knew that I loved the outdoors, and one day asked me to go along. While there Wirt asked me to help him with his herdbook. It wound up with my taking charge of that end, and, later, the breeding and public sale end.

Herefords were beginning to swing back. The range was getting much interested, and I began to see, under Mr. Armour's progressive methods in everything, that the herd had a future. So, with a penchant for publicity, I began naming the heifers with an Armour prefix, after the first year. The celebrated Armour Maids were all born in 1894, the Armour Naiads in 1895, and later came the long list of "flowers" and other feminine names, with the Armour prefix.

"Billy" Cummings, Armour's cow buyer, probably one of the best buyers in history, was in close touch with rangemen as they came to the yards, and took on the private selling end, while in heifer and cow sales I was taken along as a "pedigree-shark." Our first crop of bull calves sold at the age of about nine months, going to New Mexico at $45 per head. The second year they sold for $65 to the Prairie Cattle Co. of Colorado, but after that they began to go in small lots to various rangemen, the price jumping up along the line. But few heifers or cows were sold except from a cull standpoint; in fact, Mr. Armour began to buy, and, as I recall it, bought 25 cows with calves at foot from Gudgell & Simpson at $100 each, calves not counted, and quality the regular rotation of drop. From Jas. A. Funkhouser he bought the wonderful cow Queen Mab, bred to Hesiod 2d, at $300, and turned down May Day, then carrying Hesiod 29th, at $500. Mr. Armour never quite forgave me for deciding that she was too high in price when her calf, a year later, sold for $500. I have always felt that we made two mistakes: first, in not buying May Day, and, second, in not buying an outstanding Hesiod 2d sire, and bringing back into the herd the line of blood of The Grove 3d to combine with our Anxiety 4th strains, and The Grove 3d cows.

The days when we visited the Funkhousers are fragrant memories of wonderful drives, of an ideal host and hostess, of dinners better than kings have, and of chats with Mr. Armour, whom, next to my father, I loved better than any man. Loved? Yes, idolized, with all the strongest outpourings of the human heart where man loves man. I shall often refer to him; somewhere in this series I shall devote a chapter to him, but I must get off by myself. (My manuscript may bear the tear stains which a man wishes to hide from his fellows.)

"When the frost is on the punkin and the fodder's in the shock," when the day dawned sweet and clear, with the delicious thrift of October air, Mr. Armour would say, "Never mind what you have on hand; let's go over and see Brother Jim." We took the train to Lathrop, Mo., and usually, while I was hiring a team, Mr. Armour, a persistent trader, would buy a team of horses or a few mules. Our drive to Plattsburg, Mo., and to the Funkhouser farm, took us through a rolling, wooded country, with little streams here and there, bridged with the picturesque structures of before the war, often with the elms meeting over them. In October Nature's most wondrous brush had painted a landscape no artist could reproduce. There were hard maple reds, cottonwood and hickory yellows, poplar silvers, multi-colored pawpaws, and the deep ermine of vines clinging with suffocating ardor to giant oaks; redbirds whistling goodbyes to dying summer, and, as if in conclave for the flight to winter quarters, the little bird folk of the woods kept the music of God's great outdoors echoing the exultation of our souls.

Photo by Reeves

S. M. S. Cows in Spring Form

How we loved it all! The smell of the woods we knew would slowly give way to the odor of chicken frying and the rest of the wondrous aromas that drift in from the kitchen of country homes. Pound cake is my weakness, served warm and buttered as eaten. We knew that it would be a part of the meal. We always asked to go out and see it brown and cooling, and get a "whiff." Angel's food is all right for angels, but I am in the other class, and pound cake is good enough for me.

The Funkhouser home, a one-story brick, vine-clad and picturesque, nestled in a glorious maple grove. A distinguished trio met us, "Brother Jim," his wonderful wife and Will Willis, her brother, the well-known herdsman. "Brother Jim" had the right system: he never asked you to look at his cattle until some one had helped you up from the table. There may have been some one strong enough to turn his back on a trade after eating one of Mrs. Funkhouser's dinners, but he is not of record. She was an ideal hostess, a woman of unusual attainments, comely and entertaining, her soft Missouri southern accents suggesting ante-bellum days, with their graces and courtesy. Mr. Armour said to me, on our way home, "I believe that 'Mrs. Jim' is a better salesman that 'Brother Jim'; she gets you in the humor to pay any price he asks." He then added: "I am willing to buy a cow or a bull any time to sit down to one of her dinners." Its results may have been commercial, but I am sure that Mrs. Funkhouser's part of the work was deep-seated in the joy of hospitality.

Mr. Funkhouser had the rare gift of being able to pick a great sire as a calf. Hesiod 2d made him famous. I have heard it called luck, but I believe that most men who have developed great sires have had the "picking gift."

As the result of their initial intimacy, Mr. Armour and Mr. Funkhouser made their first joint sale in Kansas City in 1898, using the old stockyards horse auction ring. At that time Truelass from the herd of Queen Victoria topped the sale at $1,025. Lady Laurel brought $1,000. The imported heifer Lalla Rookh, a two-year-old, brought $1,000, and the bull Kansas Lad Jr. $1,000—new records after the slump of years. The 113 head averaged $385.

It was at this sale that Geo. W. Henry fell ill of pneumonia, and died at the Midland Hotel. He had been a liberal buyer at the sale, taking among others the young bull Kansas Lad Jr., a bull that we made a mistake in selling, as his after-history affirms. Our decision was made because he had practically no white on his crest, and Mr. Armour loved ideal markings. We all regarded him as otherwise the most beautiful calf dropped up to that time.

Kansas Lad Jr. was the sire of Prime Lad 108911, whose dam was Primrose 80150, brought over by Kirk B. Armour in his third importation, and sold in the same sale as Kansas Lad Jr. It will be recalled that Prime Lad, in the hands of the late Wm. S. VanNatta, became one of the greatest sires of his time. I undoubtedly made the greatest mistake of my service with Mr. Armour's breeding problem by not fighting to a finish to retain Kansas Lad Jr., but, as I have said, or will say somewhere in this series, Mr. Armour was a trader. He loved to buy as well as to sell, and never turned us down on anything we wanted in connection with our breeding problem. But when it came to a choice between a trade or breeding theory, it was apt to be a trade, and he did love to sell a good one; in fact, "Brother Jim" was to some extent tarred with the same stick. They were both dead-game sports, and took their medicine gracefully when things didn't come their way.

Mr. Funkhouser left an impress on the registered Hereford industry which was more far-reaching than that of men who have since come into prominence. I make this comment without a thought of disparaging them, but rather to emphasize the faith and persistence of men like Funkhouser, "Tom" Clark, Wm. S. VanNatta, Gudgell & Simpson and others of their class who stayed with the trade during its dark days. They remind me of the trait which always produces the great men of any industry: "The man worth while is the one who can smile when everything goes dead wrong."

CHAPTER IV

MORE ABOUT HEREFORDS AND MEN

WE HAD two permanent engagements: one for a day with the Funkhousers in October, and one with Gudgell & Simpson in May. Both were gala days, and looked forward to as bright spots. Gudgell & Simpson's homes were in the historic old town of Independence, Mo. The old courthouse still exhibited evidences of the Civil War, and the story of Joseph Smith's discovery of the tablets which formed the Mormon faith was the privilege of the oldest-timers to recall. May in Missouri means strawberries, and the varieties grown near Independence outdid the catalog illustrations. We always finished our pasture visit in time to have supper at a combination saloon and restaurant, kept by a Teuton family, which was large enough to retain all the help-wages in the family. They still pounded the steak, and it was Kirk Armour's delight to get a table near the kitchen door, which he asked them to leave open, in order to hear the old-fashioned thud of the spiked mallet. May in Missouri is also housefly time. The low-ceiled dining-room was fitted with swinging fans, decorated with tissue paper streamers, run by hand-power.

They did good work; that is, they kept the flies moving but persistent.

Mr. Armour had an intimate friend who was the glass of fashion, a fine fellow, but fastidious in every way—a man who called the head waiter to complain of the temperature of the wine served. He would never go anywhere with us because he did not like the country, and was always suspicious that we would steer him up against something queer. When we went to the German restaurant Mr. Armour would always say, "If you can get Bob to come down here and eat pounded steak, and fight flies with us, I will buy you a whole winter outfit." I never landed the outfit.

I can close my eyes and still see Mr. Armour and "Billy" Cummings sitting across the table, burdened with a mammoth pounded steak, fried potatoes, cottage cheese with chives, potato salad, wilted lettuce, a bottle of home-grown grape sour wine, and bowls of strawberries filling in the space. I must not forget several breads. Flies did not count. "Billy" remarked that he would sit down with a swarm of bees in order to get so good a meal.

My first visit to Gudgell & Simpson was in the summer of 1893, shortly before their show herd was shipped to the Chicago World's Fair. That was my first look at a real show herd, and there was a master to answer all questions. Perhaps it was a young man's earnestness, an ignorance frankly confessed, but in the many years of my visits to him and his to us, and my persistence in hunting him up at shows, putting on one side of the scales what I learned from Gov. Simpson and on the other side all that I learned from all other sources, left his side the heavier. Much of my work in Texas goes back to him. Once we were looking at a remarkable bull. I said, "Governor, how much does that heart-girth mean in the whole problem?" He stood for a moment, looking the animal all over, and then, turning, he put his hand on my shoulder and said, "Everything; but let me tell you something: the front end of a Hereford will take care of itself; that is why he will be the redeemer of the range. That heart-girth will carry him over hard times, but you take my advice and spend your life building up the hind end. I bought Anxiety 4th to do all I can in that direction. My life is slipping; yours is coming on. I hope you will use it for building up." When I come to the story of imp. Majestic, I shall refer again to this vital quotation. When I went back to the Armour herd I took hours and days and weeks and years to study and apply his thought. An old man's dream had given a young man visions which neither could think of as extending to the great range industry.

In *The Story of the Herefords* Mr. Sanders gives so concise and comprehensive an account of the Gudgell & Simpson herd that I shall not burden this article with the process by which I think the greatest herd in the world was evolved. One thing, however, is pathetic: the necessity which caused them to drop the bull Druid out of their problem. It was heroic to kill him on account of his dangerous temper, but while Gov. Simpson had little to say about it I am sure that it was the sorrow of his life, and from Druid's comparatively small get his loss to the breeding Hereford world can never be estimated and, I am almost tempted to add, nor repaired.

Another remark made by Gov. Simpson stuck. A perfect bull was being examined. I said, "Governor, do you mind going over this bull and pointing out where, if anywhere, he is defective?" Again he stood and looked for a long time; then, turning, he said, "Son, this is for you. I don't want to criticise another man's bull." Then he placed his hand on the defect, and added, "But when I hear of one absolutely perfect I am going just as far as he is to see him."

Among the stalwarts of those times was John Sparks, originally from Georgetown, Tex., and afterwards governor of Nevada, and, as I recall it, the pioneer breeder of registered Herefords in the far northwest. He was also a large breeder of stock cattle, with ranges between Elko and Reno, Nev. His herd was almost wiped out by a severe winter in the '80's. The friendship between Kirk Armour and John Sparks was one of those rare intimacies which occur between men drawn together as if by a magnet, probably the closest intimacy Mr. Armour ever had in the cattle industry. Both had absolute faith in the great ranges becoming the great eventual source of well-bred cattle, and the foundation of my faith in that result came largely through being with them much, and listening to their talks. And this reminds me of a thing which seems absurd, but in a way I still resent the advent of the auto. My mind goes back to the easy exchanges of long rides behind a team as against the difficulty of close conversation in a "jitney."

In my early days on the range, driving all day with one of the Swensons, we would cover the whole ranch problem. Now the conversations are short and jerky in a car, and we find ourselves waiting for the journey's end, and crowding into an hour what we formerly took several days to drift over.

John Sparks drew his basis for registered Herefords from several of the best Missouri herds, and was in a sense forgotten until the records of Oregon, Nevada and northern California began to tell the story of improved range breeding. Mr. Sanders records that Sparks was the largest factor in that work. In 1899 Mr. Sparks sent a wonderful draft from his Reno herd, joining Kirk B. Armour and James A. Funkhouser in a public sale in Kansas City. For many years I looked after a lot of little private business for Mr. Sparks in Kansas City, and probably the most important thing I did was to locate the supply for his annual 'possum dinner in Reno. He would wire the date, and I would set the word going for live 'possums. The Midland Hotel was Mr. Sparks' Kansas City headquarters, and all the official family loved him. The steward would set a cellar-room aside, and it became a veritable 'possum den during the period of accumulation. It is told of Mr. Sparks that he ran for governor to pardon a man who he was convinced was the victim of circumstances, and innocent.

Another prominent man of those times was Frank Rockefeller. He was a close friend of Mr. Sparks and Mr. Armour. All of them bring to my mind the love of men vital in general industry for the country, and good cattle. Looking back over the history of men who have given their time and influence to the upbuilding of the cattle industry, those who have appeared to play with it as a relaxation from their major business seem to have carried it through the dark periods, and while usually too busy to look after its details have been able to draw the right kind of men to them for that work. This brings another angle of thought: the herdsmen. Some have come into the limelight, but the great mass of faithful painstaking men in that class have been known only to one another, and a few outsiders. History means well, but the workers in the vineyard are seldom known to those who sip the wine.

CHAPTER V

"BILLY" CUMMINGS: AN ALL-ROUND CATTLEMAN

AMONG the men who stand out in my memory as all-round cattlemen, "Billy" Cummings, cowbuyer for the Armour Packing Co., holds the record. He came to the packinghouse end from the farm through a long novitiate of feeding and trading, and was probably as well liked by rangemen as any buyer ever in the yards. He was a wizard for getting "first" on trainloads of range cows. His knowledge of values was almost uncanny, since he almost invariably paid a price satisfactory to producers and with a killing test in line with killers' marginal averages. His purchases rarely showed either large profits or heavy losses. He had the rangeman's instinct for averages.

I was thrown into close association with him in the early days of the Armour registered herd. He handled the private sales; I handled the public sales and general publicity. Few men are equally good in selling and buying. "Billy" was one of the few. His methods were not those of most salesmen, who suggest or lead. He pointed out, if possible, all his wares, without making a price, watched his customers carefully for indications of their preferences, and worked to round lots, often pricing animals which they did not appear to fancy at high prices, but always making fair prices on those which they fancied. Where a selection was left to him he invariably took a little the worst of it, on the theory that no advertisement is so effective as a pleased customer. Any one who did not know the business was absolutely safe in his hands. "Billy's" wide acquaintance with rangemen and his reputation for fair dealing were vital influences in the Armour establishment. Both Kirk B. Armour and "Billy" Cummings were distinct traders, and often gave me both the headache and heartache in disposing of something I wanted to keep for breeding purposes. Mr. Armour often said, "You know, Frank, I like to buy and sell, and we will buy anything you want to keep things going, from a breeding standpoint; but you will have to get accustomed to a jolt now and then when it makes a good trade."

"Billy" liked to buy good ones. He startled the breeding world when he visited the Elmendorf herd at Kearney, Neb., in 1897 and bought the Hereford show bull St. Louis at $800, the show cow Lady Laurel at $1,000 and Dimple, a daughter of Lady Daylight, at $700. Lady Laurel, after bringing the bull calf Laurette, sold in Armour's first public sale to T. F. B. Sotham for $1,000. The bull calf Laurette by Headlight sold in the same sale to D. W.

Hart, Partridge, Kans., and was acquired by Swenson Bros. when they bought the Hart herd a few years later.

That is a story I should perhaps tell here. Mr. Hart at public and private sales bought about 20 head of Armour's cows and importations, each with a heifer calf at foot, at $300, calves not counted. "Billy" Cummings offered to take every bull calf at weaning time at $100 each for five years, as they were all in effect of Armour breeding, but Mr. Hart was dazzled by $1,000 cows, heifers and bulls and refused. Bad times came on, both as to season and price. His records were inaccurate, and just before I came to Texas C. R. Thomas, then secretary of the American Hereford Cattle Breeders' Association, asked me to go down and straighten him out. I have never seen a range herd that looked worse. Shortly after I went with the Swensons, he asked me to make him a bid on the entire herd, everything counted. The cattle were as well known to me as one's children. Without going to look at them I bid $50 per head, and took the lot, and a sorry lot it was upon the arrival of the cattle, but we applied the feed cross and did some culling; in fact, we have done some culling ever since, and the 253 breeding females, which are now in evidence in the S. M. S. registered Hereford herd, are the culled accumulation, forming perhaps the most distinct holding of the old Armour herd in existence. Last year we bought from James McNeill, Spur, Tex., some 25 registered young bulls, which came from a set of Armour cows bought by Mr. McNeill some ten years after the Hart purchase, and from which we were able to select several outstanding herd bulls.

In the meantime, we are still using from Dr. Logan's Young Beau Brummel several herd bulls which were preceded by other bulls from his herd carrying the blood of Saint Grove, a bull by St. Louis and out of a The Grove 3d cow, both bought from Armour. We bought his entire bull calf crop for two years, selecting several herd bulls for our registered herd, and on several occasions bought Gudgell & Simpson bulls and bulls by Majestic, imported by Kirk B. Armour, with the net result that we secured a distinct Armour Anxiety combination. So again my chickens have come home to roost.

When Mr. Armour bought Meadow Park, a farm south of Kansas City, from the Wornall estate, and moved everything from the Excelsior Springs farm, "Billy" Cummings moved out there, going to his home near Lawrence, Kans., for week-ends. I was married in 1899; Mrs. Hastings and I moved out during the summers, and "Billy's" reminiscences are still a joy which we go over quite often.

Meadow Park was a part of a famous battle ground during the Price raid in the Civil War. I do not remember whether that engagement was the same as the battle at Westport, where bullet holes still show in some of the old brick buildings. The Yankee plan had been to fight Price just hard enough to let him advance slowly until he reached a big bend in the Missouri River between Kansas City and my birthplace, Leavenworth, Kans., where a trap had been laid. I recall the panic into which Leavenworth was thrown. My earliest vivid recollection is of my mother packing up to flee, and putting in some of my little playthings. Our town was entrenched and the home guard stood duty every night. The Wornall road between Westport, the country club and Meadow Park was the scene of the fight, and a Kansas regiment overdid the thing, spoiling the trap by fighting Price so hard that he turned back. The old Wornall homestead, built of brick in the old-fashioned solid way, still stands about a quarter of a mile towards town from Meadow Park. It became the hospital for both sides under a primitive Red Cross.

I recall the lines by the late Senator John James Ingalls on grass: "Fields trampled with battle, saturated with blood, torn with the ruts of cannon, grow green again with grass, and carnage is forgotten."

Meadow Park carried scars over which grass, "nature's benediction," had been spread. One night a neighbor came to the farm, inviting us to a chicken fry, given for the benefit of a Confederate Monument Fund. As we chatted he remarked, "This doesn't look like the place I fought over during the Civil War." "Billy" came alive like an old fire horse. "Fought over! Were you in that battle? I belonged to the Kansas regiment that turned you fellows back." The lights faded. Mrs. Hastings and I were no longer there; the house was gone, and these two old veterans looked out over fields and lived again on a battle ground, with questions and answers flying fast. "Where were you when we came up the slope? Did they take you to the old Wornall house? We filled our canteens at the spring down in the draw." When the lull came the Blue and the Gray arose at attention, hand met hand, and stood in a silence too sacred for us to break. After the old fellow had gone, "Billy" grew reminiscent, and one of his stories was so good that I want my readers to share it.

Regiments on both sides during the Civil War did more or less foraging along the line of march. The captain of the company in the Kansas regiment, to which "Billy" Cummings belonged, had a peculiar code. He held his men down to a strict observance of decency in their foraging, but, when they got into trouble, backed them to a standstill. "Billy" said that one day on the march they noticed a farmstead at which there were forty or fifty beehives. That night, with five or six others in their company he sneaked out four mules with wagon equipment and drove back to the bee-farm. They closed the slides on eight or ten hives, loaded them in the wagon and began the drive back to camp, smacking their lips in anticipation of honey for breakfast. The bumpy road must have loosened the slides of several hives, so that all at once angry bees seemed to swarm from every side, settling on the men and sweating mules, and creating general havoc, including a classy runaway. The result was two dead mules and a demolished wagon. The boys had to make a clean breast of it to their captain, who promptly reported, "Two mules and one wagon lost in action."

"Billy" Cummings retired under the Armour pension system, after some 24 years of service. The last years of his life were hampered by some bone trouble in his legs. One foot was amputated, but he continued to work in the yards for some time, often remarking that the new foot which had been given him was better than the other. Later he lost the other foot, and some complications resulted in his death. Few men in the Kansas City yards left so many friends or had so many unbroken, satisfactory relations with salesmen and producers as "Billy." He died aged 71 years at Lawrence, Kans., in 1919.

CHAPTER VI

THE LATE MARCUS DALY

AMONG prominent men with whom I have had passing association, few have left more pleasing or lasting impressions on me than Marcus Daly, the Montana "Copper King," and one of the world's greatest horse breeders. He came into my life through a service which I was able to render him, and through his habit of paying all obligations at his earliest opportunity. I spent two months every spring with Armour brokers in the northwest and on the Pacific Coast. Montana was our great stronghold. In that state I made some personal and business intimacies which were real joys. The west everywhere is big in its manhood, ideals and instincts, but somehow Montana seemed bigger than the rest. Distance did not mean anything; the people seemed to be one big family; everyone knew everyone else, and while all had their political differences and factions they were a grand lot in their collective congeniality. I have often thought that Marcus Daly, through his prominence as a horseman, who naturally followed great race meets over the state, did much to bring them together.

The public may have a vague memory of the

Clark-Daly factions, but one had to know Montana to realize that practically the whole state had positive opinions on the subject, the Daly element undoubtedly being the more popular one. That is a story that I shall not dip into here, although it would make a thriller. I took no sides, but my lines fell in largely with the Daly contingent, since it was one of comprehensive commercial interests, and, by contact, I learned of Mr. Daly's talent for picking men as well as horses. It was a great habit with him to select men between 25 and 40 years of age, and set them up in business under their own names, with a liberal working interest. Few wealthy men I have known have done so much of this as he did. He was beloved of the working classes. He had begun with a shovel and a pick, knew the problems of the laboring class, treated workmen with outstanding fairness, and supplied every comfort and safety device for them in his mines and smelters. He practically built the wonderful town of Anaconda in Montana, where his smelters are located. The Anaconda Hotel of thirty years ago had all the "go" of New York, both as to fittings and cuisine, while *The Anaconda Standard* had its own leased wire, and carried the news of the world in line with the dailies in the great cities. Civic improvement was not overlooked.

At Hamilton, in the beautiful Bitter Root Valley, were the great horse breeding stables and pastures, including one-half and quarter-mile training tracks under roof. Horses were bred and trained in that high, rare atmosphere, giving them lung power and endurance over animals raised at lower levels. It was said of Mr. Daly that he knew the mark of every Thoroughbred racehorse in England or America. He loved the breeding and training ends, equally with the racing end, and directed his own breeding, in connection with the counsel of experts. His great ambition was to win the English Derby. I recall once that in talking with him about breeding he said, "I want to breed a Derby winner. It is like shooting at the moon, but the fellow who does not shoot has no chance of hitting." He bred some great ones, but, as I recall it, never won the Derby.

I have read many stories about Marcus Daly, which have referred to him as uneducated and uncouth. That is far from my own impression of the man. He probably had but little actual schooling, but he was rich in the education and polish that come of contact. It was my good fortune to spend a number of evenings with Mr. Daly and his friends. On several of these occasions he became reminiscent, holding his hearers spell-bound under a flow of beautiful English, punctuated with rich Irish wit, thrilling adventure, some touches of pathos, and occasional approaches to the dramatic. I recall especially his account of an early-day wagon trail, and have always regretted my not having reproduced it, because it was by all odds the best story of that period that I have ever heard. My life has thrown me much with old-timers who followed the trail to the goldfields.

I have always classed Mr. Daly among that wonderful galaxy of men who conquered the frontier, and were able to build character and education with world polish as they came along. His was indeed a charming personality; he was full of human kindness, a good friend, a good enemy, a builder, a benefactor of mankind. My most direct contact with him came in connection with one of his developing plans. He resolved to build an abattoir at Anaconda, wishing chiefly to protect the hog-raising industry, and obtain a local beef and mutton supply. On one of my western trips he sent for me and said, "I have got a packinghouse which has lost $18,000 in twelve months. I want a man to run it who can make $18,000 in twelve months. Do you know where to find him?" I replied that businessmen were shaking the bushes for that sort of fellow everywhere, but there might be one left. I put him in touch with W. N. Montgomery of St. Paul, Minn., who did the trick in eight months, and, upon Mr. Daly's death, took the plant over.

Few know of Marcus Daly the cattleman. A brief reference to his limited operations appears in Alvin H. Sanders' book *The Story of the Herefords,* but even as well as I knew of his interests I was surprised a year or so later, when he sent for me on one of my trips, and said that he was going to send Mr. Montgomery down to the Armour herd and buy a basis for a registered Hereford herd in the Madison Valley. I told him of a public sale that we were to make that fall. Montgomery was on hand, and bought a number of the best. Mr. Daly had paid his debt.

I think that Mr. Daly, before his death, gave the herd and ranch to Montgomery. I know that he regarded Montgomery highly, and desired to perpetuate his own work by leaving it in good hands. Some of my Montana friends, and undoubtedly some of Montana's politicians, may differ with me, but according to my own observations of men, who have made a strong impress upon the industries and welfare of their state, Marcus Daly is in the top-notch crowd.

As to his characteristics, a little story—one of many stories which I picked up during my Montana travels—has always appealed to me. Among Mr. Daly's stalwart admirers in the state was one "Jerry" Flannagan. As the name implies, he was a true "raw mouth Irishman." He was a conductor on the Great Northern Railroad between Anaconda and Havre. I think that he was on the first train, and hope that he is still running. It was said of "Jerry" and his opposite train companion, Frank Bingham, that they had never had a serious accident, and never a damage suit. They fought snow and high water and were sometimes on duty for 48 hours, but never lost a train. Whenever Mr. Daly's car was on "Jerry's" train it was said to be a picnic to hear those two Irish wits clash. It was even said that Mr. Daly sometimes just took the trip to cross swords with "Jerry." One of "Jerry's" most intimate friends, whose name I do not recall, but let us call it "O'Grady," had double-crossed Mr. Daly. I have said that Mr. Daly was a good friend and a good enemy. He bided his time. "Jerry," like everybody else in Montana, was a sport, as to "the ponies" in particular. On one of his trips over the line Mr. Daly handed "Jerry" $1,000 in cash, and said, " 'Jerry,' they are on to me, but I want to back Soothsayer in the Long Island meet. Place this for me, but don't bet a dollar of your own. I may be mistaken, but the odds are likely to be good, and I can afford to lose. I would not want you or your friends to lose; so don't say a word, and be sure not to tell O'Grady."

"Jerry" placed the money, and Soothsayer got the flag.

The next time "Jerry" met Mr. Daly he said, "That was a hell of a tip you gave me on Soothsayer." Mr. Daly expressed surprise, and said, "I was mistaken, 'Jerry,' but I cautioned you not to bet a dollar of your own money." Said "Jerry": "With the great Marcus Daly backing a horse, what sort of an Irishman do you think 'Jerry' Flannagan is to stay out? It's the children at home that are down to cornbread and molasses, with potatoes on Sunday." "But, 'Jerry,' I hope you did not tell any of your friends, including O'Grady."

"Sure, Mr. Daly; I have no friends left, and O'Grady has a mortgage on his store."

Several months later Mr. Daly placed $1,000 on a horse that got his nose in front and sent the proceeds to "Jerry," but the O'Grady mortgage still stuck.

I cannot close this sketch without recording one of Mr. Daly's great commercial triumphs. I refer to the D. J. Hennessey Mercantile Co. of Butte, Mont. It operated an immense department store, which carried a stock that would have been a credit to any great city. Butte was an immensely prosperous town. The concern made its own Paris importations. I recall attending a ball at Butte in the '90's at which I saw as many beautifully-gowned and jeweled women as I have ever seen at a great city function. Daly had backed the man who had the talent and let him do it in his own name.

CHAPTER VII

"THE EMBALMED BEEF SCANDAL"

I OFTEN wonder, when I pick up a newspaper and see glaring headlines featuring some new investigation of the "big five packers," whether the public realizes that it forms its opinions from the accusations, and loses sight of any vindication which may come later. "The embalmed beef" and "canned roast beef" scandal which followed the close of the Spanish-American War turned the industry upside down as few things have done. Its long investigation filled columns of the daily press for months, and yet few know that canned roast beef was restored as one of the great staples in the United States Army Commissary, and has been for twenty years. Nor did the public gather from the investigation that "embalmed" fresh beef was a myth in army rations.

For ten years before going with the Kansas City Armours I was with a wholesale grocery house in Leavenworth, Kans. The purchasing commissary for furnishing supplies to frontier posts, at which the bulk of United States troops were rationed, was at Fort Leavenworth. The grocery firm was a large contractor in commissary supplies. From my earliest business experience I was brought in contact with that work and the officers in charge. Shortly after going with the Armours I was given charge of commissary supplies in their work, both for the American and English armies. It was under my direction that the process of canning uncooked bacon was evolved, simply applying the new process of sealing tins in a vacuum as against the old process of eliminating air by the application of heat. Canned bacon was originally introduced in 1-pound and ½-pound cans for the retail trade in sliced form, and was the original of what is now packed in glass jars and seen in every retail grocery in America. When the product was announced *The Kansas City Star* published a staff story about it. Several days later we received a telegram from Maj. John Weston, purchasing commissary, stationed at San Antonio, Tex., asking for samples; upon receipt of which he wrote us that he was very much interested in the product as a part of an emergency ration which the Government had under consideration. This gave me an idea, and I did not stop until it became a great factor in the supplies of the American and English armies, the latter using millions of pounds during the Boer War in 2½ and 5-pound cans, taking it originally raw, but later processed in the cans.

Upon receipt of Maj. Weston's letter I urged the Armours to extend a cordial invitation to the commissary department of the United States Army to use their plant for any experiments that the army wished to make. This was done, an army board appointed and the original army emergency ration was worked out and manufactured in the Kansas City Armour plant. It consisted of dried beef, smoked and ground (moisture eliminated) and mixed with coarse-ground or rather cracked, parched wheat, which could either be eaten without treatment or made into a soup. To this were added three cakes of chocolate, to be eaten uncooked, or made into a drink; salt and pepper were added in individual papers. I was detailed to work with the board. I wanted bacon used for the meat part, and many officers did, but the medical division of the army fought it hard, because of the possibility of trichina, if eaten raw. In vain we argued that United States-inspected meat could be used. We conducted experiments, using heavily-infected trichina meat showing how the process of curing and smoking destroyed it. This was done by feeding it to sparrows and finding the cists undeveloped in their stomachs. Maj. Weston, however, did not give up his view of canned bacon as an eventual form for army use. He bought a carload, canned in the 5-pound size, using as nearly as possible two pieces to the can, the rind removed; this was put aboard a navy vessel and sent around the world, to test its keeping qualities in all climates.

When the Spanish-American War broke out, Maj. Weston was made acting commissary general.

The story of our absolute unpreparedness, and of our little handful of regular soldiers, is too well known and too pathetic to be rehashed here.

The foregoing may indicate my close contact with the commissary department for nearly twenty years previously to the Spanish War, during which time, among its many able officers, there were two: Gen. Weston and Gen. Alexander, who stood out with commanding force. Both were practical in commercial knowledge and instinct; both would have been among the great merchants of America, had their lives fallen in that direction. Gen. Weston was a brilliant, aggressive, do-it-yesterday type; Gen. Alexander was a calm, methodical, far-seeing man. They were perfect foils for each other. Their work threw them much together. Both were fair, exact and practical, with none of the army ego or red tape arbitrariness about them. I have never known two men in any walk of business life more delightful to deal with.

As acting commissary general, Weston shared with the quartermaster department the first great problems of the Spanish-American War. Stocks of canned corn beef were small, and the process of curing meant time. Canned roast beef, really boiled beef, required only the time for killing, cooling and boiling. Gen. Weston seized upon it as immediately available, while corn beef was being cured, and intended to use canned bacon for the major meat element, with a stew of vegetables and meat in canned form, packed in a liquor or gravy, on which he made experiments with the Franco-American soup people, all the packers and other conservers bearing carefully in mind that it must be palatable.

I was in Montana on the night of Dewey's victory. My firm wired me that Gen. Weston had asked to have me come to Washington, to help him work out the canned bacon problem. In Kansas City Kirk B. Armour joined me, and we went to Washington, to find that Gen. Weston had been supplanted by the appointment of Gen. Egan as actual commissary general, and that Weston had been sent to Tampa, Fla., in charge of the southern base. It is not my thought to criticize Gen. Egan. He was wholly blameless as to the charge of embalmed beef, but he made a bad mistake in not following Gen. Weston's recommendation as to canned bacon, as will be seen later. Everything was confusion in Washington. Egan had not had Weston's experience, nor did he have Weston's commercial instinct. I did not know any of Gen. Weston's plans, except to surmise, but in the interview Mr. Armour and I had with Gen. Egan we cautioned him about sending uncanvassed bacon into tropical countries. So many things were happening, however, that it was unheeded or overlooked.

I followed Gen. Weston to Tampa, and found that his problem was a statute specification which required that issue bacon should be from the bacon short, clear side of commerce; that he could not under the law buy and can bacon bellies, which up to that time had been the only cut used for canned bacon. He instructed me to have the regular issue cut put into from 6-pound to 14-pound cans, using as few pieces to the can as possible, but with the rind on, and as little waste as possible, sending samples to him, and duplicates to the commissary general. I had worked on the product so long that I was able to give my house by wire comprehensive details. The samples were forwarded promptly, and Gen. Weston recommended that bacon be used in canned form. I learned later that one of the vital reasons for this was that when lightering from a vessel in the surf bacon saturated with salt water becomes limp. The canned form gave the only protection, and in turn would keep and be free from maggots in tropical climates, while any fat melting from heat was available for general cooking.

Years afterward Gen. Weston told me that while in Cuba the carload of canned bacon that he had bought and sent around the world came in on the vessel and was lightered in the surf in perfect condition at a time when it was vitally needed. I think it was sent to the San Juan Hill fighters. The commissary general turned the Weston recommendation down, but just before the close of the war bought heavily in ¾-pound tins, canceling the bulk of these orders on account of peace being declared before they could be filled.

While at Tampa Weston asked me to send him several iced boxes of fresh beef, which he could set out in the sun and let spoil, in order to get a line on how far beef in that form might prove available. It was purely an experiment, the beef never having been intended to be used, and it never was used. The celebrated "embalmed beef" scandal came from a similar experiment.

In May, 1898, one Alex. B. Powell proposed to process fresh meat for the Government by purifying the germs of meats so that they would withstand the destructive elements of any climate and keep in perfect condition four to ten days. He quoted as reference the managers of various southern hotels who had used meat prepared under his process, and proposed to treat meats for ½ of a cent per pound for the Government. In the investigation which followed there was only one testimony among a mass of testimony submitted that did not agree that the quantity of food was not only abundant but of good quality. The exception was that of Dr. W. H. Daly, major and chief surgeon on the staff of Gen. Nelson A. Miles, who supported Daly in his testimony. Dr. Daly's report of Sept. 21, 1898, was as follows:

"I have the honor to report, in the interest of the service, that in several inspections made in the various camps and troopships at Tampa, Jacksonville,

Chickamauga, and Porto Rico, I found the fresh beef to be apparently preserved with secret chemicals, which destroy its natural flavor, and which I also believe to be detrimental to the health of the troops."

General Miles stated before the commission:

"There was sent to Porto Rico 337 tons of what is known as so-called refrigerated beef, which you might call embalmed beef."

Dr. Daly testified that a sample of broth taken by him from a kettle of boiling beef, on being analyzed, exhibited the characteristics of boric and salicylic acids. The testimony by Gen. Weston disposes of the charge that the beef seen by Dr. Daly at Tampa was furnished by contractors, or issued to the troops. Gen. Weston testified that it was permitted to Edwards & Powell, who were interested in a preserving process, to place a few carcasses of beef aboard the *Comal* at Tampa for a demonstration of its keeping qualities under severe tests, but that none of the meat so treated was rationed out. The report by Brig.-Gen. Charles P. Eagen of the Subsistence Department contained the following, page 151:

"Our investigation showed that rations were issued, as per published schedules, and always on hand in abundance. The department exercised great vigilance in the inspection of all articles, and obtained, so far as we can ascertain, the best quality for the price paid."

Numerous tests were also made by outside chemists, at the instigation of the commission, of both fresh and canned beef, and the reports throughout specified that no trace of preservatives was discovered.

From the foregoing it will be seen that whatever overtures were made to use preservatives in fresh meats were made by the inventors of a process direct to the Government, which, had they been adopted, would have been a contract between the inventors and the Government to treat meats received in normal condition from the packers, and in which the packers had no interest or part. The public, however, picking up the first sensational headlines reading "embalmed beef" jumped to the conclusion that it was packer doings, and it stuck for years, although in the investigation the embalmed beef charge was passed quickly, and everything centered upon the "canned roast beef scandal." Before going into that I should like to devote a minute to publicity, and how it came about.

While I was at Tampa the journalistic world was marking time. I do not suppose that a greater aggregation has ever been together since that time until the Paris treaty meet. Among them was one Whelphly, who later sprang the interview with Gen. Miles, in which the embalmed beef charge was made. Whelphly had been a staff writer on *The Kansas City Star,* and I had often come in contact with him in connection with Armour publicity, due to my fol-

Throwing a Roundup on Water, S. M. S. Ranch

lowing out a consistent plan of helping writers in general to a story in or out of my business whenever opportunity permitted. He was among the ablest writers whom I have known, and had a keen scent for news, never yellow and never a gorilla, but having written stories in our plant in connection with army rations he knew of my association with supplies, and spotted me within an hour of my arrival.

They were all waiting for the first move, and restive; there was no material for a story, and I am sure that only my long association with Whelphly saved his jumping at a conclusion story. I think that it was the most difficult interview that I have ever had. I told him that my advent meant absolutely nothing to any movement; that it was the result of two things: First, to keep in touch with things from our own standpoint; second, to get definite instructions from Gen. Weston to prepare samples of several products in various forms, which meant nothing for immediate use. I said that I had absolutely no lines on any plans that the Government might have. He took me at my word, gave the line to the journalistic fraternity, and not a single item went out of Tampa as to my appearance there, nor was there any justification for any.

I have devoted these paragraphs to Whelphly by way of vindicating his eventful article. I am sure that while he of course knew its value as a story he had seen Daly's report, and, not knowing that the meats were hung up for a test, was convinced of its correctness. I have always felt, too, that the charge that Gen. Miles had a presidential bee in his bonnet and gave the interview out for political purposes was without foundation. I knew him from my earliest boyhood at Fort Leavenworth, and have always had the deepest respect for him as a soldier and as an individual. He was a great sport. I recall that in winter when the snow fell they closed the main business street for three hours every afternoon, giving it over to racing. Gen. Miles always brought in several good horses, and was probably the most popular officer ever in command at Ft. Leavenworth.

Canned roast beef was put in by Gen. Weston because it was available quickly. It was soon found that so much of it was wanted that it took not only the available canner cattle but good cutting cattle. Canned beef usually is made from the lean parts of thin cattle. The use of fatter cattle involved more fat, which melted in hot climates. Roast beef was used as a travel ration and, with the melted fats, was unsightly; it was used excessively in camp, and was not always mixed with vegetables. Packers always let canned meats stand on the tables long enough to detect leakers, but the Government was pushing every packer in America for deliveries, and did not allow enough time to detect leakers, the contents of which quickly spoil. A carload of canned meats opened in Georgia, with a few burst leakers, would convince any normal person who stuck his nose in the car that the whole car was rotten. When the investigation came there was plenty of testimony from soldiers who had gone stale on roast beef that it was not a wholesome ration. The fault was not with the product but rather with its excessive use. During the investigation I was asked whether I had not discouraged the use of roast beef, and urged the substitution of canned bacon. I replied that I had followed Gen. Weston's thought of canned bacon and my own persistent attempts to introduce it in both the American and English armies, with great success in the latter, but at no time had I ever urged against roast beef.

I shall not attempt to follow the detail of the investigation, which resulted in a Scotch verdict, and a firm imprint in the public mind that it was unwholesome, with an automatic throw-out from commissary supplies. The startling thing which I wish to record is that while I was in the office of the commissary general, Gen. Weston, some months after the investigation, I was shown a cablegram from Gen. McArthur, then stationed in the Philippines, ordering an appreciable quantity of canned roast beef. Gen. Weston replied that, in view of public opinion, he did not feel justified in filling the order. Gen. McArthur replied that a canned, unsalted meat was vital in his work, and that he must have it. The order was filled, and canned roast beef or boiled beef came back permanently into the ration of the American Army. The only difference in the product was that instead of meat boiled in great quantities and compressed, it is packed in as few pieces to the can as possible. It is the same meat and represents the same general character of preparation, but the public does not know it. There have been no headlines of vindication and the taint on the packers still stands, except in so far as it has been forgotten.

The commissary of the United States Army began to build upon its Spanish-American War experience. The obsolete short, clear bacon side was supplanted by bacon bellies, clear of seed, put down green under inspection, examined several times in the process of curing, shrunk specially when smoked, prime in every respect, canned for warm climates. It is notable that no word of complaint has at any time been voiced in the public press as to the commissary work of the recent war, nor would this article be complete without my speaking of the wonderful response which packers, great and small, and the live stock industry at large, gave to the country's call to feed not only our own boys but the armies and peoples of the allied world. There is no regret; the whole industry from producer to packer would respond just as quickly again, but the whole industry has had more or less of the hot end ever since, and in the producer's camp we are wondering to some extent what we are going to do with all our she-stuff. The shambles form the only present outlet. The cost of production is increasing by leaps and bounds; consumption is decreasing by the same stages. Where are we "at"?

The "in-and-outers" in the industry have in all probability not fared so well as those who have stayed in, taking the lean years with the fat ones. Many men on the producing side are pretty well discouraged, and will undoubtedly get out, or reduce their operations; but there will be plenty to stick, and that will be the S. M. S. policy.

CHAPTER VIII

THE "BABY" BEEF IDEA

FOR many years the agricultural press has at times, in a casual way, credited me with being a pioneer in pushing range-bred calves to be matured as finished beeves in the corn belt at an age not exceeding twenty months. While I have devoted twenty years of my life to that work, the idea was in the main obtained from T. F. B. Sotham long before I had any thought of becoming identified with range work, and while many feeders undoubtedly were trying it in a small way, even before Mr. Sotham took it up, he was the first to get behind it in any definite way. While it was sure eventually to come, his initial work brought its first impetus, introducing it as a distinct phase of the feeding industry years before its natural evolution could have been brought about.

When I began my work with the Armour herd "Tom" Sotham was a "live wire" in extending the use of registered Hereford bulls into the range. He made many visits to range herds, came in contact with the big and little men of the range, recognized the rapid improvement that had been made, initiated some experiments as to the outcome and became convinced that the market for registered bulls would increase directly in proportion to the benefits that breeders received from their use. My attention was first called to his work when he made an effort to induce a number of ranchmen to contribute to a public sale 100 good steer calves, or, say, two carloads each, to be placed by him, and developed into "baby beef." He had every natural condition against him. Tradition said that finished beef under three years lacked flavor; feeders were skeptical about blackleg, ticks, the cost of development, brands and comparative market prices as against older cattle; there was no export demand for that class; acclimation was a deterrent factor, and, in fact, from every angle it was a leap into the dark.

Producers naturally wanted to send their best, and they knew the curse which is put on a topped herd, no matter how small the cut. They had misgivings as to the net result from the sale of calves as against their spring clearances of yearlings to the northwest. They were in the habit of selling their drop, less a 10 per cent cut, to one buyer, and did not cotton to dividing up among several small buyers or selling at public sale at some eastern center. Trails to the railroad were often 50 to 100 miles, and according to the old ranch code that meant trailing the cows to the railroad with their calves, and the cows back home. "Tom" was "up against it" on both sides; progress was slow. The story is too long for all the details.

Without really knowing it I brought Sotham's idea with me in the back of my head to Texas. A year here developed it, and I sprang it on the Swensons, who gave their consent, and in the second year we brought all of our steer calves, born by June 1, to Stamford in late October, classified them, put them in feedlots on a maintenance ration, composed of cottonseed cake, cottonseed hulls, sorghum hay, and black-strap molasses, properly balanced and mixed with mathematical precision by machinery, and sold them as buyers wanted them, any time between November 1 and April 1, adding the cost of maintenance to them; which, in the first year, was $1.50 per month. At that time it was thought that cottonseed meal, except in a limited quantity, would kill a calf.

I proposed to my people that we test it out and kill 100 heifer calves by getting them up to a full-feed, and feeding them until the following May. The experiment was on; we got up to as much as 4 pounds of meal per day, 2 pounds of molasses, and all the hulls and sorghum hay (about half-and-half) that they would eat. On about May 1 we sold the entire 100 head, fat, without any evidence of "meal evil," netting fairly well on them. We went out to the cornbelt with the broad statement that well-bred Texas calves were good enough for any cornbelt feedlot. Many of our fellow producers joined in the plan. In the meantime the Department of Agriculture had been working on the economies of baby beef, giving it much encouragement; feeders were experimenting with good results; the cornbelt was becoming interested; pilgrimages were being made, and mail inquiry was becoming enormous, but few producers were willing to sell less than their drop. We were probably the advance guard selling any considerable number, one or more cars, as wanted. This was made possible by concentration, classification, and straight cut-offs, which has since been changed, to the extent of selling in advance, and shipping all of one class from one ranch, at the same time, first grading to a standard and loading by a straight cut-off.

We soon found that our difficulty was in getting feeders who wanted only one or two cars to come down. This in turn evolved the mail order idea, one which obtains to a much larger degree over the whole range country than is generally known. There were many headaches and some heartaches in the pioneer work. I recall that John Camp, Harristown, Ill., a feeder himself, and acting as agent for us, brought down a number of his neighbors. We had concentrated 3,000 steer calves and 1,200 yearling steers in our pens. His men were slow, and walked and walked, looking at the cattle in different pens. Mr. Camp, E. P. Swenson and I sat on a fence waiting for them to get through. The humor of the work struck me, and I said: "John, I dreamed last night that conditions had changed; some fifty cornbelt feeders were here, and I said, 'Line up, gentlemen, for your turn.' The first man up and asked the price, and whether I did not think it a little high; to which I replied, 'Get down, and give the next man a chance.' Then I woke up with a cold sweat all over me."

Many years after, during a season in which we did not have half enough calves to go around, Mr. Camp wrote me: "Your dream has about come true."

I shall not attempt to follow our own public sales with C. C. Judy at Tallula, Ill., or "Tom" Sotham's attempt to build up a great public sale business at Kankakee, Ill., except to speak with pride of "Tom's" great come-back in recent years in handling public sales of registered cattle, and one incident in our own work which gave me the final stimulus to go on, in the face of difficulties.

We announced a public sale in carlots for February 1, in 1905, at Tallula, Ill. Every one thought that I was crazy to ship cattle in mid-winter for public sale, but probably no one knew that we were getting short of water in our feed pens, which were supplied from the city supply, and were forced to do something. We started 1,000 calves and yearlings in a bad storm; they reached Tallula when there was an 18-inch snow, on the level. Two days before the sale the mercury dropped to 20 degrees below

Separating Cows and Calves at Shipping Time, S. M. S. Ranch

zero. We sold the cattle in a tent, when the weather was 10 degrees below, shipped them over a wide radius, did not lose a single head, and every carload made a good market record that summer or fall. After that I was sure that Texas had the goods to stand any climate, and it was just a matter of staying with the enterprise.

A rather amusing incident occurred while we were at Tallula. I had taken two outstanding cowboys with me, men who simply had to know what was to be done, and did it in spite of hell and high water. They had been up there with me in the fall, when the red apples hung on the trees, and the weather was fine. They loved it, but with 20 degrees below zero weather we cut holes in ice on ponds and drove the cattle out, so that their weight would flood the surface.

We were spelling each other cutting through the two feet of ice, and I was resting, after my turn, when the humor of the situation struck me, and I began a rhapsody on the beauty of the north. "Boys," I said, "I was born in the north. I love these snow-clad hills, this bounty of ice and the splendid invigoration of zero temperature." For some minutes I did a good job in a grandiloquent way. Finally one of them rested on his axe, and, turning with a look of supreme disgust, said, "You can take your damned north and go to hell with it; I wish I was there with you to warm up some."

In the years during which the International classes have dropped out the three-year-olds and the championships gone to yearlings until a great feeder like John G. Imboden asserts that there should be two championships, one for two's and one for yearlings, because the two's no longer have a look-in; when the yearling markets are on the average higher than for the heavier classes; when we no longer read that flavor only comes with age; when the Department of Agriculture is insistent that the economic production of finished beef is in the yearling class, it would seem that calves at weaning time, furnished in great numbers from the ranges, have come to stay. My own part in the change has been much smaller than I am often credited with; it has simply been the logical result of a study made by feeders, abetted by the splendid work of agricultural colleges, and the research work of the United States Department of Agriculture.

CHAPTER IX

SOME JOURNALISTS I HAVE KNOWN

I OFTEN think that the law of contact has more to do than any other factor with our lives. According to some notable opinions coming out of our colleges the value of erudition is outweighed by the human factor of contact with young men and women from every part of America, each contributing some developing influence. In my own life the look-back over the journalists whom I have met seems to be the greatest asset in my human savings bank. Father Bigelow was the editor of *The Notre Dame Scholastic* in 1876, while I was attending Notre Dame University. He told me that I had some inherent qualifications for journalism, but that I was rottenly crude, being too much inclined to bombast and sentiment, and needing a fine-tooth comb, which he presented in the way of reading a chapter from Addison's *Spectator* every day, and reproducing it from memory the next day. I wish that I had been more persistent, but it was like doing the "setting-up" exercises every morning—good things, but how many stick to them?

After I came out of Ann Arbor I loafed my off-duty hours about newspaper offices, and rather drifted among newspapermen, some of whom have filled, or are filling, high places. I have, however, made it a consistent part of my life never to look up men who have grown great. William Allen White was a cub reporter on *The Kansas City Star* in my 30's, and I treasure in my scrapbook a generous review by him of some work I did. George Horace Lorimer, the brilliant editor of *The Saturday Evening Post,* was with the Chicago Armours while I was with the Kansas City house. We were cordial friends. The following lines by him have probably passed out of his memory, but, after all, greatness is only a discovery, and while I have been told that John Hay was ashamed of his *Pike County Ballads,* it has never seemed possible that the man who wrote *Jim Bludso* or *Little Breeches* could possibly wish to disown them. Mr. Lorimer may not feel the same way about his verse, because while he rose to greatness in his book *The Letters of a Self-Made Merchant to His Son,* in which he has drawn a wonderful picture of the late Philip D. Armour's sound basic business philosophy, he may shy at being credited with doggerel. We were getting out the Christmas edition of our weekly price list. It was elaborately decorated with holly, Christmas bells and steaming plum pudding, with a New Year's poem (doggerel) from my pen. The excuse for reproducing the three stanzas is to permit a comparison of it with Mr. Lorimer's clever parody:

A NEW YEAR'S THOUGHT.

A moan and the old year passes away;
  A smile and the new one is born;
A world of hope for the coming day,
  And a sigh for the one that is gone.

Here's a cup of joy for every home,
  With hearts full of happy love;
Another glass for those who roam,
  And a blessing for all from above.

Here's a word of regard and memories kind
  For our patrons far and near:
A voice of thanks on every wind,
  And for all a Happy New Year.

The reply follows:

A NEW YEAR'S THOUGHT.

(After the Kansas City Bard.)

A grunt and the old hog passes away,
  And along on the hooks he's borne,
And it's sausage he'll be on the coming day,
  And hams and sides and brawn.

Here's a pail of lard for every home,
  And a kit of pickled feet,
And regular tripe or honeycomb
  And a thousand things to eat.

Here's our business card and wholesale list
  For our patrons far and near:
Don't go to Missouri to get your grist—
  You can buy it cheaper here.

—The Sweet Singer of the Chicago Stock Yards.

The only other notice my verse received was from an Arizona editor, who said:

> "The Armours are out with a very touching New Year's poem to their customers, most of whom would have appreciated more a reduction of ¼ of a cent per pound on hams and bacon."

The late Wm. R. Nelson, familiarly known as "Baron Bill Nelson of Brush Creek," owner and editor of *The Kansas City Star,* was the most remarkable personality that I have ever known in the newspaper world. He and Kirk B. Armour were close friends. Mr. Nelson often accompanied us on our herd inspections; in fact, he was a neighbor, having developed, just beyond Meadow Park Farm, a beautiful tract, stocked with registered Shorthorns, of which breed he was a life-long and consistent champion. The improvements in the way of barns, granaries and the like were the best that I have ever seen. It was not a mere plaything, as his constructive work with Shorthorns at Sni-A-Bar demonstrates. A personality which could develop, in my memory, in a moderate-sized inland city, an evening paper from a struggling initiative to one of the recognized best dailies in the United States, naturally left lasting impressions on me, even during my limited association with him. He was a picker of men. I wish that I had the space in which to review the brilliant galaxy which functioned in his office.

I obtained my first realization of the rapidly growing importance of the Argentine cattle industry from comments which Mr. Nelson and Mr. Armour made after seeing a ship's cargo unloaded in England about the time of the Paris exposition, and before England put the ban on live cattle for immediate slaughter. They were discussing it after their return, when Mr. Armour turned to me and said, "We think we are doing wonderful things on our ranges in the improvement of cattle, but, mark my word, we have a real competitor springing up in South America, and he is pushing us close now." Mr. Nelson put in a crosscut as to the preponderance of Shorthorn which Argentina was using, and yet both spoke of the excellent whiteface steers in the shipment.

It was my thought in this sketch to review agricultural writers, and while I could drift on for pages concerning those near and dear, as well as brilliant writers, with whom a casual association has occurred, I have drifted as much perhaps as space will permit.

As I come to the men who have lived in my real world, I sit for a moment with the ink drying on my pen, trying to decide how I shall attempt to write of one to whose memory every lover of God's great outdoors rises and stands at attention. There is a little monument to that memory before which I want to stand sometime before I die, and breathe a prayer of gratitude to God who sends men of that sort to work in the vineyard.

Dear old "Joe" Wing!* There is a soft place in my heart as I reach my word to his memory, and the tear which stains the page seems sacred to an association which was all beauty and joy. He sang like the mocking bird that I hear all day long as I drive over vast areas. Always on the topmost bough, his heart flowing through his wonderful throat into a never-ending song of love of the world and praise to the Master. It is difficult to say anything of "Joe" which has not been well said. He was a dreamer of dreams, with the courage and industry to make them come true, and when his time came his train of mourners comprehended the whole agricultural world, on which his impress has been deeper and more permanent than that of any other writer who loved the fields, herds and flocks, and punctuated his stories with a love of mankind.

In the earliest days of my connection with the cattle industry I came in contact with a modest, retiring personality that I soon discovered was a tower of strength and concentrated ability, whose application to study and work as an editor for more than thirty years, and whose writings on breeds, as well as abstract thought, have instructed, guided and inspired every live stock breeder. Alvin H. Sanders has done so much constructive work outside of his journalistic

*The late Joseph E. Wing of Ohio was for many years and until his death in 1915 a staff correspondent of *The Breeder's Gazette.*

profession that an attempt to review it, except by mere touches, would be to repeat what breeders and stockmen are familiar with. My comments will therefore be confined to personal reminiscences, but I cannot pass on without expressing the pride that I take in the wonderful literature, both technical and philosophic, which he has given us. I recall recently having sent his *In Winter Quarters* to my daughter in an eastern college, with the suggestion that it be given to her instructor in English as a classic of western literature, and a gem of descriptive and philosophic English.

When Herefords began to come back in the early '90's the late Kirk B. Armour was better known individually among rangemen than any other man, not actually in the range industry. When in town he always rode, a part of each market day, with his buyers in the Kansas City yards. His advice as to the policy of improvement was sought daily. His verbal and public print universal statement was, "I believe that only the best results can come from introducing registered bulls, regardless of what beef breed the buyer may select." The Hereford association wanted to elect him president. Mr. Sanders, anticipating that the members would do so, came to me to get Mr. Armour's photograph. The Armours did not believe in that sort of publicity, but I did, and, knowing how I would be "landed on," I furnished the photograph, which was reproduced on the front cover page of *The Breeder's Gazette.* The storm broke, as I had expected. I had never known Mr. Armour to be so provoked as when he stood at my desk and said: "This is some of your work, and I do not like it." Next S. B. Armour "landed," and then Herman O. Armour, who happened to be out from New York, took a shot, and finally Philip D. Armour gave me a "once-over." All did a good job. To all I made the reply that the name Armour was on goods in most foodshops in America; that Kirk Armour was personally the public sponsor for cattle improvement, and had been honored with a great office; that it was good general publicity to bring the name pleasantly before the public; that they could fire me, but that I would never lose an opportunity to do the kind of thing that I had done. Of course, it all blew over, in good nature, but it was years before I could get Kirk Armour to "stand hitched." Mr. Sanders and he were great friends. Mr. Armour valued his opinion higher than that of anyone in or out of the industry.

During the Paris exposition Mr. Sanders became very much interested in Limousine French cattle, and in Normandie cattle of the milking strains. He wrote us his impressions. Mr. Armour instructed me to write to him to buy a string of both, and send them to us, and added: "We shall buy some of the milking Shorthorns of England, some Ayrshires from Scotland, and some Jerseys and Guernseys. Have them all tested before shipment, and they must be tested in quarantine, giving us a certified dairy herd; then we will start a dairy on the Jewett Farm with up-to-date appointments, have our milkers disinfected and sell certified milk."

It was only one of the many progressive things that he wanted to do, but with his declining health I had instructions to find some reason why they could not be done. I shall not burden this sketch with particulars concerning what I did. It was a great sorrow to have to find reasons for not doing things, when I wanted him to do them all. I could fill pages with accounts of the delightful visits that I have had with Mr. Sanders on farms, when looking over cattle, and chats in his office and at shows and sales which have been my privilege with the gifted editor, and it would probably surprise his modest estimate of himself if he knew how much of his wisdom that I have utilized.

The late W. R. Goodwin, associate editor of *The Breeder's Gazette,* received so many deserved eulogies following his death that I shall only pause to add my personal opinion of his great work in making *The Breeder's Gazette* an acknowledged leader of the world's agricultural publications. My last interview with him occurred when he was preparing his great editorial review of the world war beef situation. He outlined much of it, and I carried away the deepest impression of how carefully and analytically great questions were studied and weighed by him before they found expression. In over thirty years of our close association, when I wanted a profound opinion I sought him. I recall that when in Florida, in the winter of 1917, I met him casually in a restaurant in Jacksonville. His and my time was limited, but in talking an hour over the studies that I had been making, everything that I wrote later about Florida carried his impress.

Among the bright men whom I saw much of in the early public sale days was Geo. P. Belows, an unusually handsome man, of strong personal magnetism, grace of manner and decisive, courteous speech. It was in every way logical that he should drift into the live stock auctioneering business, in which, but for his untimely accidental death, I am confident that he would have become the premier, and yet the agricultural press lost a great exponent when he changed his vocation.

The late Kirkland B. Armour, head of the house of Armour & Co., Kansas City

CHAPTER X

KIRK B. ARMOUR AS I KNEW HIM

MANY references in this series have been made to the late Kirk B. Armour. They are interwoven with his association with men and events in preceding chapters. I shall therefore try to make this sketch resemble an intimate personal portrait. I came in contact with him by chance in 1889. It was one of those chances which confirm the thought that while it is every man's province to figure his future, as one would do a sum, the real influences in our lives usually come from the most unexpected sources. This statement sounds fatalistic, but it is not so intended. Senator Ingalls wrote that "opportunity knocks but once and, passing, knocks no more"; but he was mistaken. It is knocking all the time; and intelligence, intuition, or free moral agency decides what opportunity offers.

I had gone into the Kansas City Armour's office with a Government inspector, who had some business there, while I had none. As I sat waiting, practically without realizing why it was done, I walked into the main office, and asked if the house needed anyone on the trade-getting side. I then came for the first time in contact with a personality which has had a dominant influence on my life. Kirk Armour stood 6 feet, perhaps a trifle over; he was built in proportion; a massive chest, broad shoulders, a square jaw and laughing eyes. His face and hands were bronzed by the great outdoors that he loved. Little waves of magnetism seemed to radiate from him, even before he spoke, and when he spoke a mental arm seemed to come up to salute. In that initial interview I became aware of qualities in him which I have heard commented upon hundreds of times; as, for example, his interest in individuals, his love of progress, and his doing things worth while. He asked questions so rapidly that they could hardly be answered, and before I knew it I was hired. I took away with me the same impression that everyone took who ever talked with him ten minutes: "This man is interested in me." It was true of all; he was interested in them and their work; he had suggested something progressive. It was his interest in everyone, development, progress, betterment in everything, which proved a burden even greater than his Herculean strength could carry. It sent him back to Mother Earth at the age of 48. I still mourn some of the things that he had planned; things which would have been great community benefits, often of national importance, if his life had been spared. I have often felt that he would have become the world's greatest merchant.

Of all the incidents that I can recall which illustrate his gentleness of heart, the one of the great packer—"the daily spiller of tons of blood"—and the humble rabbit seems best. When we went to the Excelsior Springs farm on Saturdays, Watson and Lawrence Armour (Kirk's sons, then mere children) usually went with us, getting all the sport that boys should out of a farm visit. Once, while driving in the old spring wagon through the pastures, a young cottontail ran out from its nest. The boys were out in a minute, and we all joined in the pursuit, sprawling about in the grass until it was caught. The boys were of course for taking it home for a pet. Mr. Armour, holding the frightened little fellow in his hand, turned to his farm manager, "Charley" Wirt, and asked, "Do rabbits cause much damage?" "Charley" said, "Yes; they gnaw the young fruit trees, and do lots of damage to the garden." Mr. Armour stroked it for a moment, put it down in the grass, and, as it scampered off, said, "Well, I guess this one will not make much difference." I have always loved Bret Harte's poem *Luke,* and this came into my mind—

"And she looked me right in the eye—I'd seen sunthin' like it before
When I hunted a wounded doe to the edge o' the Clear Lake shore,
And I had my knee on its neck, and I jist was raisin' my knife
When it give me a look like that, and—well, it got off with its life."

Mr. Armour came into the business very young. He was put through all the various branches under outstanding men who had instructions to give him the "third degree"; but they worshipped him, as all did, and the proudest boast of every one of these men, when Mr. Armour had become the master, was, "I taught Kirk that." He spent part of each work day in the saddle at the yards, and always looked over some part of the plant with the superintendent or some department manager, calling by name great numbers of men, and stopping to chat with them, often about their family affairs. His memory for details was remarkable. I have often heard him remark, as we were discussing something, "You had different views some years ago," sometimes quoting in detail. I recall, when we were out in the plant, making a special investigation at the ham-testing table, his saying to the ham-tryer, "George, how is the leg of that boy of yours doing?" Some days afterwards I was at the same table, and asked George about the boy. He replied, "Oh, that was three years ago, but Kirk never forgets anything."

Mr. Armour loved trade. He did not bother us much in the winter months on the provision side of the house, because it was the dull time, and we were all sent out on trade survey trips for an average of six weeks; but when things began to open up in the spring he was on our backs continuously. It was an old saying in the office that a man who did not get roasted at least once a day never got his salary raised. I recall an illustration of his conviction that trade would always expand under effort, and had no limit. I had been given charge of a product which had only a moderate sale, and by taking advantage of some favorable natural conditions had an unusual streak of luck with it. A card showing the manufacture and movement of the product in every class was placed on Mr. Armour's desk each morning. A glance over it revealed to him the weak spots. In glancing at my product card he got the wrong line; that is, another product showing, say, 90,000 pounds per day, while mine showed 180,000. He strode over to my desk, and said, "What's the matter with you? Going into a decline? We ought to be selling 100,000 pounds a day." To which I replied, "Hell, boss, we are selling 180,000 pounds," but my raise of 80,000 pounds never touched him, and, without batting an eye, he said, "That is not enough; we should be selling 250,000 pounds. Shake yourself; you are walking in your sleep."

He always reminded me of a little negro boy who told me of thinking that he saw a ghost on his way home one night. Describing it in his own language, he said: "When I come up on that ghost I lit out as hard as I could, and every time my feet hit the ground I says to myself, 'Ed, you can do better than this.' "

Mr. Armour was an inspiration to every man under

him. All caught his trade-getting instinct, and took all the hurdles. It was fast company. He had that most wonderful of all business faculties: the faculty of eliminating jealousies and building team-work. Behind it all there was a personal love, almost worship, and loyalty to the last ditch. In the twelve years during which I was under him I never heard a single employee criticise him. I meet men all over the United States still in the business or in other fields, and it is always with reverence that his name is mentioned by them. Sometimes a number of us have chanced to foregather and exchange recollections of the old days. There will be some individual anecdote or beautiful memory; a voice will grow husky; there will be a halting of speech, a choke, and the story goes unfinished. Eyes will fill, someone will brush his cheek or hastily arise to get a glass of water; then we talk of other things. But out of the grave of it all comes the sweet peace of memories, golden threads woven into life's more somber raiment.

Some twenty-five years ago an effort was made to consolidate the great range industry into a cattle trust. I do not recall its exact nature, but Mr. Armour stood out at that time as having a broad, personal acquaintance with rangemen, and I shall always think that no man has been so vital in his relation to the improvement of cattle or the extension of meat products sales round the world. One afternoon we left the office to drive to Meadow Park Farm. Mr. Armour stopped in front of the old Midland Hotel, and told me to hold the team while he went up to a meeting. When he came down he told me the nature of the meeting, and added, "They offered me a million dollars in stock for the use of my name, but I told them that I did not want to make a million dollars that way."

I never knew Mr. Armour to be severe but once. I shall not go into the details. Briefly, a man sitting in the executive office, in charge of a department, stole $45,000, and never handled a dollar. It was of course a case of clever outside connivance, and one of the things that led up to departmental bookkeeping. We all knew that the man was living beyond his means, but that was accounted for cleverly. It was during the days of open gambling; there were faro lay-outs everywhere. Often after he had got his "divvy" he would show, to enough people to get the word about, a wad of big bills, and say, "I hit them hard again." An accident led to our identifying the thief, but instead of prosecuting him Mr. Armour made him handle his desk for six months, isolated, despised, a Benedict Arnold, an outcast. Once he said to a former intimate friend, "I would much rather have gone to the pen for ten years."

Mr. Armour loved horses and the country. A byroad in the wood always brought from him, "Let's see where this goes." The wilder and rougher it was the greater his joy. He would stop to listen to a brook or pluck some wild flower for his buttonhole. No matter what flower it was, wild or cultivated, he would always look at it a moment and say, "That is the most beautiful flower that grows." Once with my wife I was in Chicago on New Year's Day. We went out to get what she loved best, a single American Beauty rose. The florist said that one could not be found in town, and asked, "Do you know that they are worth $30 to $40 a dozen?" I asked why. He said that farmers all over the country had sent in so many Christmas and New Year's orders that everything was cleaned up at enormous prices. I told the story to Mr. Armour. In a flash he said, "Let us build greenhouses out at the farm. We go out there now and shiver around in the winter looking at the cattle, but with a flower business we could enjoy the cattle in the summer and the flowers in winter." His health, however, was failing fast, and I had to find a delay for not going ahead.

He was a remarkable combination of gentleness, aggressiveness and public-spiritedness, with much of the far-sightedness of his uncle Philip D., by whom he was loved as a son. He was easily Kansas City's greatest and most beloved citizen. The mourners who stood, tear-stained, as the last words were said, "Dust to dust," numbered as many of the lowly as of the high, and in their hearts it was pure gold lying there under the flower-covered mound. One of them who had worked with him, and bossed him as a boy on the old New York State farm stood at my side. I felt his hand grip mine as the others turned to go, and heard him say, "Wait." Then, when we were alone, he crossed himself as he knelt, and sobbed a broken prayer, and I—well, that's all.

CHAPTER XI

BRAHMIN CATTLE; CULLING A RANGE HERD

WE have never used any Brahmin blood, but it has a wide use in Texas as far north as the Texas & Pacific Railroad. I do not know of any Brahmin cattle in the Panhandle or in central-west Texas, except steers, brought from south Texas for development. Here it may be interesting to record that a few years ago a large northwestern steer buyer gave instructions to eliminate anything showing the Brahmin cross, but before he did so some had gone through. Several years later, when these few had matured, and gone grass-fat to market, he changed his instructions, taking everything with the Brahmin cross on a par with the general offering, because he saw that the samples had made fine weights, and were extra fat, nor were they discriminated against in the market.

The history of the Brahmins may be briefly given as that of the humped "sacred" cattle of India, imported by the late "Shanghai" Pierce. The development of their use was made under the direction of A. P. Borden on the Pierce estate, near Corpus Christi in Texas. Very few purebreds were brought over. The process of amalgamation has been mainly through 1/2, 3/4 and 7/8-Brahmin crossed with all the beef breeds but with Herefords predominant and Shorthorns next.

The breeding process has been so varied and technical that I shall not attempt to dissect it any further than to say that it is a widespread interest in southern Texas, and has some of the best breeders in the state for its practical champions, with results which cannot be ridiculed. My attention was attracted to this work by the operations of E. C. Lassater, "Al" McFadden, "Tom" East and James Callan, all in Texas. The judgment and ability of these men as beef producers place them in the top row. All that I know about the cross has come from these men and from a visit to the McFadden Ranch, where I saw many crosses and results, both from a breeding and beefmaking angle. These to me proved to be a revelation. There I met a man who had always "cussed" the breed. I asked him why he did so, and what were their limitations. I got a knockout answer. He said that the Brahmins had no limitations, from a moneymaking standpoint; and that he just naturally hated them, but was going to breed them from now on anyhow. In proportion to the blood used, they are immune from ticks, flies and other insect pests; they will go further for water, lie out longer in the sun, when other breeds seek shade, live on coarser grass and weeds, get rolling fat earlier, and kill out better than our native cattle.

A cowpuncher who worked on a ranch as far north as Dryden, Tex., where a good deal of the Brahmin-Hereford cross obtained, in talking of the early shipments from that section in the spring of 1920, when grass-fat yearling heifers and steers were netting from $50 to $65, told me that the stock carrying Brahmin blood was the biggest and fattest, and then added, "We were short of water, but those scoundrels just got up and trotted ten miles for it, and got fatter all the time."

People who have learned how to handle them do not have any trouble, but a little thing like a fence or a corral does not seem to bother them at all. Mr. Lassater asked me to locate a cornbelt feeder who would full-feed a load of one-fourth-Brahmin and three-fourths-Hereford calves, against a load of S. M. S. calves, both lots to be billed at the same price but under a protection of $10 per head that the Brahmins would net as much money. I found someone, but he only kept them a short time. Even then, however, they made him a small profit without the $10 protection, which was of course only extended for a full-feed. I have always regretted that the experiment did not go to a conclusion. I do not think that the Brahmins will invade the herds of our section and the Panhandle as breeders for some time, but I do think that they will keep on, and that some day cornbelt feeders will use the best crosses extensively in their feedlots.

One distinct peculiarity of these cattle must be recorded. The bull will serve the same cow only once; he is very sure and rivals the goat in capacity. The quality of the meat, it is said, holds its own against that of any of the beef breeds in comparative finish. I can vouch for that, as I ate some from a good, grass-fat carcass served by Mr. McFadden when the executive committee of the Cattle Raisers' Association of Texas met at Victoria, in June, 1919. I am told that there is no discrimination in the market against Brahmin cattle; in fact, at Fort Worth, where the Brahmin cross is heavily marketed, it finds favor because of its good kill.

While in Florida in 1917 I found Brahmin bulls of strong grade being used on the run-down primitive cows, with the thought of getting scale, constitution and the beef instinct. It was the intention of crossing the heifer progeny with the established beef breeds. In theory it looks good to me, since many of the Florida grasses are coarse, and, as I told them in an address, Florida cattlemen should try anything and everything, since they can't breed down.

The law of selection must be the natural basis for bringing a herd to the highest production and quality, but all breeders will agree, I think, that individual merit must stand severe rivalry from "the get." I remember that in discussing breeding with Marcus Daly the matter of "like begets like" came up. He said, "That may work in cattle, but it makes some strange misses in horse breeding." Then he fell into a dissertation on maternal influence which I wish I could reproduce, because "maternal influence" is my pet hobby.

It is impossible to follow the bull in range breeding. All we can do is to work largely toward individual merit. We cut 35 per cent as yearlings, and thereafter as often as development suggests. Bulls can be bought, but a great cow herd only comes from accumulation, persistently culled with "get" as the prime factor. I recall that incidental to correspondence in reference to the Armour Hereford importations in the '90's the late W. E. Britten, who selected them for us, wrote that a celebrated English breeder had a cow named Lady Fickle, which carried one of the richest pedigrees in the English herdbook, but was not much herself. He added, "But she has the blood, can be bought reasonable, and has a good bull calf at foot by a good bull." We instructed him to send her along. I met the shipment at quarantine, and told "Bill" Searle, who brought them over, that I wanted to see Lady Fickle first, and he remarked, "You will not see much." She was indeed a terror, cat-hammed and flat-ribbed, with sprawling horns and a long dished face, but very broad between the eyes, and carried what John Gosling called "a brain-box." Her calf was fine, and sold for a good price. Mr. Armour did not want her around. She sold as

Photo by Erwin E. Smith

Cutting out Cattle by Classes, S. M. S. Ranch

a cull with another bull calf at foot which developed into an outstanding show bull and sire. On the other hand, we bought Beau Real's Maid, one of the outstanding show cows of her time, paying $2,200 for her. She dropped a heifer calf by a great imported bull, brought over by the late Chas. S. Cross. I have rarely seen a poorer calf, and it sold at a trash price in a general clean-up to go some distance. I was never able to follow its progeny. So there you are: two jolts, one on each side.

We follow both plans on the S. M. S. Ranch, and try to save the good individual with a good get. Our culling is done in the fall, when the cow with a calf at foot may be seen, but invariably we leave some cows which we do not like but which produce calves in the top row. It is almost uncanny the way in which foremen or cowpunchers generally will speak of an individual cow among thousands of cows as having had several good or bad calves. Without attempting to explain this fact, I have checked against it carefully enough to know that they know what they are talking about.

In culling cows, threes and over, we always try to work to type. We throw out a cow, no matter how good, if she goes too long without having a calf, or appears to be a persistent misser—another class which the men spot.

Nourishment is watched carefully. Poor "doers" in the winter are thrown into a small pasture in the spring, as are many other classes, subject to a second look-over in the fall for cows with one eye, spoiled bags or other physical defects. Age of course is a factor to which we give careful attention, but it does not eliminate a cow if she is a "doer" and a "repeater" with no definite limit. A really ill-looking cow is rarely kept. A rather old cow with an outstanding calf is apt to be given another year as against one of the same age with even an average calf.

I was asked once where I placed "get" in the culling problem. I replied first, because it is the final test in realization. We rarely cull anything in the calf period except distinct ne'er-do-wells. All calves from two-year-old heifers go out, as it is impossible to avoid having some yearling heifers get in calf, and all calves at foot with culled cows not old enough to winter go. An average term of years will find about 5 per cent of the total drop going as veal for these causes.

The main cut in the breeding herd is made in the fall, in the yearling heifer class. It is about 10 per cent. This is followed in the two-year-old period by cutting anything missed as a yearling or not making proper development, which brings the herd up to the three-year-old or over class, when they take pot luck in an annual cull, running rarely less than 7 nor more than 10 per cent. In steers the cut, running through from calves to yearlings, will average about

10 per cent, leaning toward 12 per cent, but varying to some extent with the season, and veal shipments not considered.

The methods of different ranchmen will naturally vary, and the remarks that I have made are only intended to cover our own methods, which will apply, with small variations, to the general Texas industry, with this exception: southern Texas frequently has all classes fat early in the spring, when there is no rule except the protection of maintenance by holding back such top she-stuff as is needed, and marketing the classes returning the best results. The year 1920 has shown the fattest cattle from southern Texas in years. A good clearance has been made, including heavy hold-overs from 1919.

CHAPTER XII

SOME WESTERN CHARACTERS

BORN in the historic old town of Leavenworth, Kans., in 1860, I have vague memories of what I now know as the Civil War, which, to a child's mind, meant only hanging over the gate to watch soldiers march by or fill canteens at our well, to hear the bugle call, or the hushed voice of my mother and her friends as they speculated on the fortunes of war and the threatened "Price's raid," or, as Bret Harte puts it in *Miss Blanche Says*—

"Still it was stupid: Rat-a-tat-tat! Those were the sounds
of a battle summer,
Till the earth seemed a parchment round and flat,
And every footfall the tap of a drummer."

Kansas was abolition; Missouri, just across the Missouri River, was Confederate and largely guerrilla. Col. Jennison's celebrated abolition cavalry regiment was stationed at Fort Leavenworth. His men were "wildcats," every one a picked horseman and consummate daredevil. It was said of that regiment that its men were on the guerrilla order and especially careless about the title of a horse; in fact, the usual pedigree of a horse was "by Jennison out of

Missouri." Years afterward, as "Dan Quinn's old cattlemen" put it, the colonel reformed and started a saloon, with a faro bank attachment. His place became a famous resort for old-timers, and as it was more respectable in those days than now to drop into such places, I often listened with mouth agape to stories and reminiscences which would make a wonderful history of the part "bleeding Kansas" played in the Civil War, and the gateway to the west, which Leavenworth formed for the years following.

Majors Russell and Waddell, with their hundreds of ox teams and wagon trains transported overland freight to all parts of the west. Scouts, guides, hunters, trappers, desperadoes and a great stream of fortune-seekers to the great unknown poured through the gateway. Buffalo hunters marketed their hides, which filled great warehouses. My grandfather shot a wild turkey on our back fence. "Kit" Carson, "Wild Bill" and "Buffalo Bill" (Col. Wm. F. Cody) were as familiar figures on the streets as ordinary citizens. The Missouri River was another trail to the great west, and was alive with fine boats carrying their burden of human and commercial freight to Fort Benton, Mont., the head of navigation, and from that point freight was scattered by bull teams over Montana and Utah.

Durfee & Peck, Indian traders, with frontier posts, had their headquarters at Leavenworth. I still have some of the trade brass money which they issued for furs, and only the other day in Chicago, as I looked into a Michigan Avenue shop window, I ventured to ask the price of a wonderful mink garment. My mind harkened back to the absurd price at which the skins could be bought in the early days. John Bowles, at the Chicago stockyards, wore a beaver coat for which some friend had paid $1,500 and loaned him to wear while he auctioned off John Hubly's grand champion load of Aberdeen-Angus steers at the 1919 International show. "Tom" Todd, Fort Benton, Mont., showed me a coat forty years ago made from selected Saskatchewan beaverskins which cost him, including the making, $58. I slept out in the snow between Great Falls and Belt, Mont., in a bewildering storm thirty years ago wrapped up in a beaver coat, which the driver offered to sell me the next day for $90. Still, look where sugar has gone in two years! Perhaps furs are still cheap!

"Buffalo Bill" was born in Salt Creek Valley, 5 miles from Leavenworth. Romance has made him almost everything that the west, in its wildest woolliness, is capable of; but as a matter of fact his real business was that of supplying the Union Pacific Railroad building crews and camps with meat, which was mainly buffalo, and from which he took the name known all over America and Europe. Naturally this pursuit gave him endless adventures. I think I have said before that I never look up notables, but when "Buffalo Bill" came to Stamford with a circus some time before the illness which resulted in his death, my boy, who had heard me talk of Mr. Cody's early life, and had read numerous stories about him, asked me to take a little bunch of boys to call on him after the afternoon show.

I introduced myself from the Leavenworth standpoint. Col. Cody exclaimed, "Why, I know your father well! I bought guns and ammunition from him; he used to sell the old Dupont powder." Then he had us all come into the wonderful shelter tent kept for him between performances. The boys will never forget that visit, and Col. Cody's kindly chat with them. The hand that shook "Buffalo Bill's" is still a tradition in Stamford. I sought to go when the boys went, but the colonel detained me for an hour or so, asking about all the old friends whom he had known in the early days, many of them dead, and as those who had gone beyond were mentioned a shadow of sadness would flit across his face, followed perhaps by some story of such men as Levi Wilson, Len Smith, Capt. Tough and Alexander Caldwell—men of force and personality in the '50's and '60's.

Col. Cody then made me tell him all I knew of the Texas cattle industry, and in turn he told me about his own ranch near Cody, Wyo., adding, "You know I have been over this wonderful west until it is just like going from one room in your own home to another, and when I had seen it all I picked out the one outstanding spot. I believe it is the best ranch property in America."

We talked of the "Leavenworth man," and the bunch that scattered when the railroads came, and moved on, with frontier restlessness, and then he told me a remarkable thing: that while his wild west show was in Paris during the exposition he made it a habit to try to locate a "Leavenworth man," and have him ride in the stage during the fight between the cowboys and Indians. He said that it was remarkable how often he found the man. When at last he let me go he said, "Well, this has been a wonderful hour through you, living again with my friends in those happy times. God, but they were happy, happy times! Look me up any time you can. I am hungry for more of the long ago."

I recall one striking thing that he said when, inadvertently, in speaking of the value of a buffalo's hide at the time of our chat, and of the slaughter for their hides, he said: "People have the idea that I was a crack shot, and used that talent to kill for the hide. I never killed except for food, when the hide followed as an auxiliary perquisite. I never have had any patience with the vandals of the plains."

"Buffalo Bill" was not a "bad man," but in the early days he was sometimes confused with "Wild Bill," who was on that order. He was just a hunter, scout and guide, who in after life was clever enough to turn his remarkably picturesque figure, personality and talents into a fortune, which he did not know how to keep, but in the making of which he has left in the hearts of two continents the romance of the prairies, now rapidly dying under the advance of the hoe. Col. Cody had a heart as big as the story his great wild west told. History will record him as its most picturesque frontier type. Steeped as I am in the love of the frontier, I read *The Literary Digest's* compilation of American press tributes to Col. Cody after his death with the feeling that they were strewing flowers on something of my own—the wonderful west.

Capt. W. S. Tough was one of the outstanding—I might almost say romantic—characters of those stirring times, since the story of his life would make fiction tame. He stood over six feet, built in proportion, a born horseman, and a dead shot. He was in the secret service of the Government during the Civil War and a United States Marshal immediately following, when the reaction from war and the great flow westward brought him into action with outlaws and bad men generally. He became one of the famous peace officers of the border through his cool, picturesque courage. I saw much of him as a boy on his breeding farm next to my father's, and later (in 1880) in Denver, where he owned a horse sale barn, making a specialty of single and team roadsters, much in demand at that time. I was lonely, and he invited me to spend my spare time exercising his horses on the Denver speedway, under instructions not to let anything pass me. He had a genius for mating teams with striking effects, producing matched, mismatched, showy gaits and nobby turnouts. All his drivers dressed the part. When he drove himself—and he often drove—it was worth going to see. He was the best driver I have ever known. Often when we spent the evening in his office, he would grow reminiscent. I recall two of his stories which may be of interest.

During the war he was on secret service on some important mission in Missouri, near Bee Creek. He was captured by a band of guerrillas. It was nearly dusk when they reached camp. He was turned over to a drumhead court-martial. The trial lasted a few minutes; the verdict was "Spy; penalty, death." Turning to him, the leader said, "Yank, would you rather be shot before or after supper?" To which the captain replied, "I always did think a lot of my belly," adding, "Say, 'Reb,' that dun hoss of yours is the best I ever saw. Suppose you make it sunrise; I'd like to die looking at him." "All right," said the "Reb"; "come eat." At supper the talk drifted onto horses, and the quality of the Missouri horse, then, as now, their pride. "Reb," said the captain, "if you ever miss that dun hoss you can just figure my spirit has got astride and gone with him."

They tied the captain to a log, with the "Reb" on guard; the fire burned down to the faint glimmer of coals through ashes, and he fell asleep, awakening to find his hands and feet free and the sentry nodding over the smoldering fire. He crawled behind a bush, took off his shoes, carrying them in his hand, and made his get-away. Ten minutes later a shot rang out in the woods, followed shortly by more shots. When morning came he hid for the day. The next morning found him at the Missouri River, which he swam to the Kansas side. After the war the captain hunted up the "Reb," gave him a mighty good horse and big boot for the dun. "Yank," said "Reb," "that hoss talk of yours did the work; I thought it was a damned shame to shoot a lad who loved a hoss like you did, and, after all, you was only doing your duty." As he ended his story, the captain said: "They call me 'nutty' about a horse, but every time I look at my wife and babies I think, after all, a hoss gave them to me, and, next to them, a hoss comes first."

While commanding a band of cavalry scouts—naturally all picked horsemen, and crack shots—they took dinner, paying for it, at the house of a woman whose husband was in the Confederate army. The captain had given his men strict orders not to forage, except in cases of extreme necessity. A magnificent mare of the distinct Missouri saddle-horse type, and about six years old, was grazing in a small paddock near the house. The captain and all the men admired her, and "Scotty," a young daredevil and sharpshooter, tried to make a trade for her; but the captain told them all, "Hands off." They camped for the night some 20 miles away. Shortly after daylight, "Scotty" rode in on the mare. As their eyes met he said, "Well, Cap, I left a damned good hoss for her, and she's contraband, anyway." "Scotty," the captain said, "my orders are not contraband; get your breakfast, and ride back. We will wait for you here; and, Bill, you go back with him."

"Scotty" replied, "I'll be damned if I do."

"All right," said the captain. "One of us has got to be boss, and we can't be that with the other around. Ride to the other edge of this clearing, turn and I will ride towards you; shoot whenever you get ready. I will do the same. You are at least as good as I am on the trigger."

"Scotty" sat for a minute in deep thought, and then, with a smile, said, "O hell, Cap, there ain't no doubt about who's boss here. I only traded for this mare to keep you from getting her after the war. So long; catch up with you by noon. I'll get one good ride out of her." Then a bright thought striking him, he added, "Say, Cap, you ride her back; maybe you can make an after-war trade for her." His story ended, the captain was silent for a moment, and then broke out into a hearty laugh. " 'Scotty' was right; you have heard or known of the good ones I bought and sold for twenty years after the war. Well, I spotted them or their dams and sires during the war, and what Jennison didn't get during the war I tried to get afterwards."

Another story may be added because it falls so aptly in Missouri. During President Cleveland's first campaign I was sent over to Platte City, Mo., with several others to organize a regiment of cavalry to participate in a big night rally, and parade in Leavenworth. Those were the days of the Flambeau Clubs. All the "pep" has gone out of parades since the use of fireworks has been prohibited. We had three of them in line, a total of 200 expert Flambeau men whirling pin wheels, and putting up rockets or Roman candles as they marched. Nothing that I have ever seen compares with that sort of display.

We chose a day when Gov. Thomas T. Crittenden was billed to speak. Jesse James had been killed by "Bob" Ford under a contract with Gov. Crittenden. The James boys had worlds of friends in Platte county. Gov. Crittenden had been warned that he could not speak at Platte City without being killed. They were all there, and I asked the Governor to give me a minute, before he spoke, to announce our cavalry regiment, which, by the way, came 500 strong, on probably as grand a bunch of horses of that number as ever formed a regiment. When the Governor rose to speak there was a tense silence; not a single hand clapped; the dropping of a pin could have been heard. He was calm and cool, as though shaking hands with a friend. Those who knew him will remember his splendid presence, his wonderful snow-white hair, his square jaw and his well-modulated voice. For a moment he stood silent, his eyes roving over the entire audience. Everyone seemed to be holding his breath, and then the Governor began:

"Fellow citizens of Missouri: I am your Governor. I am pledged to enforce your laws. I have been warned that if I touch here today the subject of the James Boys and the killing of Jesse James I will pay for it with my life. I am here to talk to you on that subject, to tell you the whole story, and my final attitude, and then drift to national and state issues. Bandit-ridden Missouri, your Missouri, is the scorn of America. Your laws charge me with responsibility. I am the guardian of Jesse James' wife and children, and I intend to keep that trust to the last ditch; but I did put a price on Jesse James' head; I will put a price on the head of every bandit and outlaw in Missouri, dead or alive, so long as I am your Governor. I stand for law and order, and I will follow relentlessly all violators until the fair name of bandit-ridden Missouri is back where you can stand in any company, and say, 'My Missouri,' without a blush, and if I did not have the moral courage to come to you and tell you so I would not be fit to be your Governor."

Then he paused. Again an awful stillness ensued;

it seemed to burn and hurt. Then, as though shot from the mouth of a hidden cannon, the people rose to him en masse, and cheered him to the echo. A great, tall, grizzly-headed man, sitting in the front row, walked up on the platform and took the Governor by the hand, waving for silence with the other. Like the shutting off of an electric current, it was again still. The old man said: "By God, boys, we've got a Governor. Let's stay with him." Bedlam thereupon broke loose.

Then followed the regular speech, but probably the shortest the Governor ever made—the once-over on national issues and an appeal to send a real Platte county regiment to the rally. That first few minutes in its dramatic tenseness had made everyone weak, and the good old applejack, for which Platte county was (and perhaps still is) famous, seemed a beautiful refuge. I know it was.

CHAPTER XIII

SOME COWBOY CHARACTERISTICS

I SHALL write of cowboys as I found them in the spring of 1902, when I came to the S.M.S. Ranch in Texas. There were whole outfits of trained, seasoned men, almost born in the saddle. This situation obtained pretty well up to the war, when voluntary enlistment and the draft took the cream of cowboy material, leaving only the foremen and straw bosses (second in command) to train the material that ranchmen could find. Since then we have robbed the cradle and old men's homes with an ever-shifting outfit, built up round a few old-timers. Yet it has been wonderful how ranchmen have got along. It is explained by the fact that the really skilled old-timers are beyond the age for new ventures, can still handle the technical work, and have infinite patience with raw material.

After the war many of the cowpunchers came back, but were restless; then a matrimonial epidemic swept most of them into wanting camp jobs, and there were not enough to go round. A working outfit must, in the main, be unmarried; otherwise it is "busted" half the time by normal and legitimate demands to call the boys home. Do not let these com-

Bringing Back a Truant, S. M. S. Ranch

ments bring an avalanche of applications from agricultural schools or lads generally wishing to learn the cow business. I shall not in this sketch elaborate the reason why we must draw from local men who have lived in the atmosphere through boyhood. There is of course a reason.

There is beginning to be a swing back from the oilfields, the harvest fields and town jobs. Only a few days ago a well-dressed city-complexioned young man came into our office and applied for a "riding job," meaning in the cow outfit, as a fence or pasture rider, as opposed to farm work. I said to him: "Pardon me, but you don't look the part. What experience have you had?" Then he told me that he had been hurt by a falling horse about a year ago, and been working "inside" until his leg got well. I told him that we did not pay "inside wages," but he got his keep. "I know that all right," he said, "but when I get through paying board and room rent, buying cold drinks, good clothes and taking the girls to the picture shows, wearing a clean shirt every day, getting a bath and 'dolled up' at the barber shop, and paying for the gas when a friend takes me for a joy ride, me and the world is several bucks apart at the end of the week, and feeling like hell, too; but I can save money, eat good grub, ride good hosses, shave myself, get one of the boys to cut my hair, and take a runnin' jump into the creek or tank from the brandin' pen; look up at the stars as I fall asleep out there on the ranch, and I'll stay hired a long time if I get a chance." He got it.

After all, he had it right: it is what you have at the end of the month. Perhaps the reaction in favor of farm and ranch may come through that channel; there can be no question about the health end.

Cowboys are ultra-sensitive, diffident and superstitious about anything that they do not understand. I do not mean the supernatural; I mean things out of their own groove. They possess a quality that is not necessarily courage, but rather the absence of fear. They are not the lawless, gun-toting element of the movies, or the long-ago frontier. They are law-abiding, good, hard-working citizens, with a greater respect for the chastity of a good woman than any other class of men in America. I have studied them for twenty years. I have asked hundreds of cowmen and cowboys if they knew of a single case of seduction, and have yet to hear of one. I saw a line in one of the films recently disclosing that some easterner had followed a girl whom he was in love with to a ranch, where she was visiting. Finding his case hopeless, he began to shine up to a country girl. The girl whom he had followed had caught the spirit of the country. She said to him: "Ned, the things men do where we came from and are dismissed as wild oats, they kill for out here." A country mother once said to me: "I would rather trust my girl with one of these boys any distance alone in a car or buggy than have her go three blocks alone in a city."

When I take a bunch of cowboys with me to a hotel or café, where the service is "classy," I do not think of asking them what they will have, but look over the bill of fare, and pick out what I think that they like, ordering it for myself, and then, turning to them, ask, "What will you boys have?" "Same for me." They will go back home and laugh about the mistakes they made using the wrong silver, and drinking out of finger bowls, but if one of them were conscious of attracting attention he would die of mortification.

As to superstition, an incident occurred in my first year which will best illustrate it. We had carried over some cattle, and had an unusual number to deliver, trailing them mainly to the Tongue River Ranch, 100 miles distant, and thence 60 miles to Estelline. I was in the field constantly, and often stayed with the outfits weeks at a time. I had said to them when I came, "Boys, I am soft, and I do not know much about this big pasture game, but I do know something about breeding and selling. Take me on faith until I make good, and let's do teamwork. Some day I will ride as fast and as far as any of you." It was a go, but they couldn't help having their fun. I got some pretty rough-gaited horses and on the round was often sent with some crazy rider. After an especially hard ride over rough country one morning, with an especially crazy rider, I decided to stay at the chuck wagon, instead of going on the afternoon round.

That evening after supper, around the campfire, the chuck wagon "josh" commenced. There is nothing funnier or sharper. I came in for the brunt of it. Finally I said: "Boys, I told you I had lots to learn about this riding game, but there are some things I learned before I came down here. I am a trained athlete (those who know my stature will be amused). I will wrestle any one of you, catch-as-catch-can, for $50. Now that does not mean to stand up and take hold, but run in, catch by the head, or leg, or any way. It is all sleight or trick work, and that is where the little man may put it over the big one—the comparatively weak man over the strong one, just as the 'Japs,' trained in jiu-jutsu, can handle several men not trained in strangle holds, arm twists and stomach jams. So in catch-as-catch-can the man with the sleight may hurt someone who does not know it, and I want that understood."

I had been taught a few tricks, and, reaching over, caught a big fellow quickly by the back of his neck, jerking it down, and almost unbalancing him, while the others laughed. The "bluff" went; they had never heard of catch-as-catch-can, and were all skilled in the sleight of throwing a calf well enough to know that some of the smaller men could beat the big ones. They looked foolish; then one said, "You try him."

I saw I had them going, and remarked, "Pool your money, and pick your man," but it was the unknown, the "sleight," and we wound up by putting it off, but they quit "joshing" my riding.

Some months later an inspector whom I had got to know and love asked me, "Where in the hell did you get to be such a wrestler?" I had really forgot all about it, and asked, "Why?" Then he told me that my boys were offering to bet the other outfits that I could throw any man they would put up catch-as-catch-can, or, as they put it, "You got to wrassel his way; it's all sleight and if anybudy gets hurt he has to grin it out."

I never had to wrestle, but some years afterwards I ran a footrace with a lad, who was just a kid "hoss wrangler" at the time of my "bluff," and he beat me. Thereupon, throwing his hat down, he said: "Now, damn your soul, I'll jes wrassel you catch-as-catch-can." I told him that he had been off somewhere taking lessons, and I did not want to hurt a good man, anyway, because they were getting too scarce. To which he replied, "Say, I believe you've been runnin' a sandy all the time." I said, "O. K., kid, but keep it dark."

I have said that cowboys are diffident and unemotional; that is, on the surface. You can do things for them, and while scrupulously polite in the etiquette of the range they may not even say "thank you"; but they never forget. Under that nonchalant exterior there is a heart of gold; a tenderness and response to every human touch. The cowboy is indeed a child of nature, and all its great emotions are in his soul. I had this all come to me in the saddest experience of my life when one of our boys was killed while we were sorting cattle in the feed-pens at Stamford. He was riding his night horse Curlew. A night horse is always the gentlest in a man's mount, but any ranch-bred Texas horse will go out of his head when anything goes wrong, often some very little thing.

I remember riding this same horse on an all-day 50-mile jaunt over the Tongue River Ranch with Henry Bonner of Indiana. The day was very warm. I took my handkerchief out to put round my neck, and gave it a flip. Curlew went into the air, and for a minute jolted me pretty hard, and made me pull leather. I said to the foreman, "What's wrong with this horse?" He replied, "I guess he never saw a white handkerchief before."

"Johnnie" was sitting on Curlew, and several other boys were on their horses, waiting for instructions. I think that Curlew must have fallen asleep. The foreman rode up and said, "Come on, boys." Johnnie was one of our best riders; he was always "cowboying." Reaching over he gave Curlew a slap on the neck. The horse must have jumped in the opposite direction. Johnnie was unbalanced and fell, his foot catching in the stirrup. The horse then frightened, made several plunges, striking Johnnie in the head with his hoof. The boy's imprisoned foot came clear, and in a few minutes Johnnie came to, and seemed quite himself; but I sent him to camp near by with one of the men, who soon sent back word that Johnnie was acting queer. I sent a man "riding for the doctor." That means *riding*. All work was called off. We gave first aid, but it was evident that a blood clot had formed. We put in a long-distance call for the Pitch Fork Ranch, 80 miles away; it sent a rider to the boy's father and mother 20 miles away. It was then 3 o'clock in the afternoon; relay teams were arranged along the line.

I knew the frontier dread of the knife. The doctor told us that there was just one chance: trepanning the skull. I called a council of the boys, and said, "Now if he were mine I would take the chance; but I know how men on the frontier have seen their comrades lie for days and weeks unconscious, and get well, their rugged constitutions finally absorbing blood clots, and they have seen death follow heroic surgery. Dick (his father) should call us from Aspermont (40 miles away) by 2 o'clock in the morning. Shall we wait?" There was no answer; then I polled them one by one. "Wait" was the unanimous verdict.

At 2:30 a. m. Dick called. I told him just what I had told them. Poor Dick! The agony in that calm, restrained voice—"Wait till mama and me come, Frank. I've seen so many of 'em in my fifty years lay and lay and come through. Wait."

I went over to my house across the street for a short rest. The boys had followed me about like children; they would squat on their heels along the hall while I was in the sickroom, and when I came out there was that mute appeal, with never a word, but just the anguish of soul, crying for a word, a look of hope. I told "Mage" to call me if there was any change, and asked the doctor not to leave the room.

I was just sinking into the sleep of exhaustion when I heard Mage knock on the open window, and heard him say, "Frank, come! Johnnie is about to go." Then we all gathered about the bedside, a band of stoics, not a tear from hearts bursting with grief, and Johnnie passed into the Great Beyond. Gentle women were there to meet the father and mother when they came, two hours later, and the boys, standing their silent guard of honor, moved out into the hall, leaving them alone with their dead.

We arranged for starting back with the remains on the following day, but along towards evening rain began to fall. There were two sand rivers to cross which were swollen torrents under a head rise, and it might come down by morning. So it was decided to start at once. Night was falling; there was no time for a service, but the mother said, "I want my boy to have a prayer"; and I lifted up my voice to thank God for the refuge that He gives to all in the great range of love and peace. Then, with Johnnie's riderless horse, saddled and bridled, with a slicker-clad, slouched-hat guard of honor in front, behind and on each side of the spring wagon, which held all that was mortal, a weird and solemn procession moved out into the slow drizzle, and gathering darkness, on its 100-mile journey to the little spot on the hill where the first sunbeams kiss the morning dew, and the flickering shadows of dusk linger longest. There Johnnie dreams in the eternal sleep—

"Where the little grey hawk hangs aloof in the air,
And the sly coyote glides here and there;
Where the summer rains and the winter snows,
Where the wild flowers bloom and the Bob White goes."

CHAPTER XIV

CATTLE WEIGHT COMPARISONS; TICKS

I AM asked to make a comparison of the weights of the Longhorn and his more modern brother, raised under improved breeding. This is a task which I fear can not be satisfactorily performed; first, because the comparative ages of marketing varies too widely, and, second, because the data of early-day weights are meager. I have found in my reading only one definite reference to weights (this was in about 1870), and that was unaccompanied by any data as to age or sex. It recorded a drove of 1,200 head, original Longhorns, cut from 3,700 head, and showing at the market 1,125 pounds. The probability is that they were steers, and cut for weights and from general data they were presumably from four to six years old. I could perhaps get data from some old-timers, but they are scarce, and would require to work largely from memory. I shall therefore work from another angle.

In my own memory the ages at which cattle are marketed have shown drastic changes. When the International Live Stock Exposition in Chicago began twenty years ago, three-year-old steers were in the favored class; export buyers were taking what are known now as "heavies." Then twos were the next step towards youth in the favored class, and finally yearlings. I have never worked in the Chicago yards, but venture to say that cattle from the northwest developing ranges have in the meantime shown a material drift in age. When I came to Texas twenty years ago there were many threes; now, under normal conditions, they are rare, except in the extreme Panhandle, which has become more of a developing than a breeding country. A well-posted stockyards commission man once threw a bombshell into my camp of theories by saying that improved breeding was being done at the expense of weight, citing a well-known brand with which he was especially familiar; but in going into the matter deeper I was convinced that the marketing age had been reduced, and that a year's keep cut some figure. I do not mean to say that he opposed improved breeding; his point was that the price must be greater to offset the loss in weight. I followed the matter further into the question of what percentage of increase in choice cuts resulted, with satisfactory testimony that it was material. My final conclusion was that improved breeding results in materially earlier maturity, and a saving of from one to two years' keep. I have never been convinced that actual losses in weight, year against year, have been against the improved animal.

An illustration of much interest came under my

own observation. The Spur Syndicate trailed through our Tongue River Ranch on the way to Estelline. Its cattle were among the first in the state materially to be bred up. When I came to Texas one of the Spur steers had "outlawed" in our pasture several years before. He was thrown into every Spur herd which passed through for several years; but he was a professional outlaw, and was back on his range near the T-41 Windmill in our pasture in a day or so. He must have been a two or three, when he first located on us, and was still there five years after I came. He looked like an elephant beside the other cattle. Finally I had him thrown into one of our own herds, got him to Estelline, and shipped him for the Spur account.

As I recall it Fred Hosbrough told me that he weighed between 1,800 and 1,900 pounds, off grass; at any rate, he was very heavy. At another time I sent one of Burk Burnett's stray steers in. I think that he was a five-year-old; he showed a very heavy weight. I have talked with a number of rangemen who have had similar experiences, bringing up the matter of comparative weights, and in every instance they have shared my conviction that improved Texas cattle of today, age for age, will put the primitive cattle to sleep in weight. My own observations are purely from Texas experiences. A better comparison would be the cattle shipped from the northwest ranges to the Chicago market, which has absorbed the supply for forty years or more. I should like to see *The Breeder's Gazette* put "Jim" Poole on this job, to dig out data from the memories and records of old-timers in the Chicago yards, still in the trade. It must be remembered, however, that in the early days of the Great Northern Railroad the late James J. Hill spent a great deal of money putting out Aberdeen-Angus bulls along that line, and quite a bit of breeding was done in Montana. The Aberdeen-Angus bulls were crossed on cows probably pretty well graded. The comparison should probably be made against Longhorns taken to the northwest in the '80's and cattle in recent years, with cattle in both instances fat.

The foregoing has been written more with a view to "starting something" and bringing out information from all sources than to claim any merit for my own observations. The topic should be thoroughly discussed by rangemen and cornbelt cattle feeders and breeders.

It is quite a jump from weights to ticks, and yet the extended work of the Federal Bureau of Animal Industry must have had a great influence on weights when the drain of tick vampires on the early cattle is considered. I am sure that immunity from ticks within a few weeks after trail droves started north had much to do with their weights in Montana. I have often heard old trail drivers speak of starting with droves of thin cattle, taking them from south

Texas by easy stages to Montana, and arriving fat. They were naturally free from ticks in a few weeks. When I came to Texas, Chicago had established a free season, when cattle from below the line were admitted to the regular yards, without prejudice, as the result of definite knowledge that ticks dropped from November on would succumb to northern winters. The same practice applied to many northern states, which admitted them in the open season without inspection.

I have followed the quarantine map carefully during the past twenty years, the last ten of which have shown amazing changes. For five years after 1902 we had ticks in the S. M. S. Throckmorton pasture; but we dipped and cleaned it several years before the pastures across the road were clean. We drove our calves to Stamford every fall, dipping them for exposure, as they had to pass through infested territory. We did not lose a calf from infection, and only a limited number from dipping—not twenty head, all told, out of thousands dipped, and most of those few got turned around in the vat and drowned. Crude oil was the only dip that we ever shied at; other dips apparently were an actual benefit, while crude oil seemed to close up the pores in the hides.

Robert Kleberg, manager of the King Ranch, was first among Texas cowmen to tie onto the tick theory, and his constructive work in the early period brought benefits which can never be estimated. For years it was a moot question in the Texas legislature, which has in recent years got behind the work of eradication with both feet, coöperating with the Bureau of Animal Industry splendidly, both by precept and appropriation. There was from the earliest discussion one argument which never failed. Get a man who rejected the tick theory to bring cattle into infested territory, which had never been exposed to ticks. The result was always the same, with this one exception: a fever-tick which found a horse for host dropped and laid its eggs, and the progeny hatched did not carry the fever germ. Some pastures as far north as King and Motley counties became infested by ticks dropped from horses, and there was no bovine mortality until they got onto cows brought from below the line but clean when brought in. From such cows the tick again took on her malignant nature, and her progeny would cause fever. In these northern counties, however, ticks rarely carried over, on account of the cold, but a story made up of scares would make a book, and I had my share of them.

The quarantine line now, according to a Federal bureau map (Dec. 1, 1918), is 90 miles south of Stamford and the same east, with no unclean territory north, and due to the work of the past two years the line is much further south. In five years, with proper local and state coöperation, the fever tick could easily be cleaned out in America. I feel reasonably sure that it will be in ten years, or, if not, then at least limited to portions of Florida. That state, however, is getting a move on itself, and if the fencing of open range there is rapid enough, it can clean up easier than Texas. Some wonderful work has been done in Florida by progressive individuals, and the state has passed laws in which there has been some confusion in interpretation.

The United States Department of Agriculture in Washington, D. C., has issued a number of comprehensive pamphlets on ticks and tick eradication. From these publications a few extracts, condensed and unquoted, and some observations and data from Chief Dr. John R. Mohler of the Bureau of Animal Industry, and from Geo. M. Rommel, chief of the Division of Animal Husbandry, may prove valuable to those interested.

Texas fever was introduced into the United States with importations by Spaniards during the early colonization of Mexico and southern United States. It seems to have been first discovered as a menace by Dr. J. Pease, towards the close of the 18th century in Lancaster Co., Pa., where a severe outbreak of the disease occurred following the importation of some cattle from North Carolina. Pease ascribed the cause directly to the cattle, without locating the tick as the cause. Experience soon showed the invariable result of the importation of southern cattle in great mortality among northern cattle. Years followed before any tangible cause could be located. Smith of the Bureau of Animal Industry in 1889 was the first to locate the tick as the carrying cause, while Kilboune of the same bureau in 1889-1890, by conclusive field experiments, suggested by Chief Salmon, proved the presence of the tick to be essential in the transmission of the fever. In 1891 Dr. Salmon established the Texas fever quarantine line, but for ten years he had been working on the theory, and was probably its actual discoverer.

The Texas fever tick is continually confused with a number of other ticks, occasionally found on cattle, which, so far as the transmission of Texas fever is concerned, are entirely harmless, and which one of the department's earliest problems was to define and demonstrate as harmless. Farmers' Bulletin 569 by Dr. John R. Mohler is the most comprehensive pamphlet on the subject that I have read, and an historical address by Dr. U. G. Houck of the Federal Bureau of Animal Industry is probably the best thing of its kind.

The ear tick, while harmless as to fever, is a menace at times in the northern parts of the Shinnery country of Texas. Coal oil or gasoline, injected by hand into the beast's ear, is the only practical remedy. Ordinary cattle dips do not penetrate far enough. Fever ticks claim the short-haired parts, and are rarely found on the head, and never in the ears. Often in winter the ear ticks will form a perfect cement in a cow's ear. We keep a lookout for cows of this class, which are usually in low condition, finding their way to the feed grounds. One year ear ticks were so bad that we put the whole feeding bunch through a chute, and squirted coal oil into their ears, with good results.

In closing this chapter I am constrained to pay a tribute to the splendid work done by the Federal Bureau of Animal Industry, through trying circumstances and against much opposition, in the matter of obtaining individual and state coöperation. Its painstaking, patient methods are saving the world many times the cost of the whole bureau every year. Eventually, due to the bureau's initiative and persistent, constructive work, America will be free from fever ticks.

The late G. F. Swift, founder of the house of Swift & Co.

CHAPTER XV

SOME PACKERS I HAVE KNOWN

THROUGHOUT this series my reference to packers has been largely to the Kansas City Armours, with whom I spent twelve years. During that time I came and have since come into contact incidentally with the most of the other packers. It is my thought to review them from the standpoint of their influence in rescuing the cattle industry from chaos in the late '60's. I shall limit myself to those who have passed over the Great Divide.

At the time I went into the service the packing industry was emerging from a comparatively small business to the spread-out stage. Refrigeration was being improved by leaps and bounds. Branch houses, making the local distribution of fresh meat at great distances possible, were increasing rapidly; circuit cars were in their infancy, and men were being sent into every part of the world on trips of survey. Swift & Co. had recently built in Kansas, but were killing cattle only. One of my most thrilling memories is that of creating new by-products, following up the investigations of our scouts, and of the intense rivalry among the men on the sales end, with packers to put over a new one or get general business.

Packers were large intertraders on long or short lines, as the flow of trade or specialization suggested. For instance, my house had an immense trade in pigs' feet. We bought the rough product, uncleaned, from several packers, who did not clean a hoof, but sold just as few in finished form as possible, and to fill orders bought them in tierces from us, filling smaller packages under their own labels. This intertrading gave us all an inter-acquaintance. Some of my warmest friends today were strong trade rivals, while my work in Herefords took me much about the yards, and in close contact with packinghouse general yards buyers, and in turn I came at times into contact with the high bosses.

I once heard "Johnny" Bowles at an International Live Stock Exposition auction sale speak of Nelson Morris as "that grand old Trojan." I think it covers my own estimate of him. My own association with him was casual, but he knew me whenever we met, and from the earliest days of the International always had a kindly word and "How are those S. M. S.'s coming along? You got some good cattle, and you got a good boss." It happened that when the senior member of our firm, E. P. Swenson, was coming to Texas for ranch visits, Mr. Morris would be making a similar trip to see his black cattle near Midland. It was a wonderful herd, and its dispersion left grand footprints in many parts of Texas. Quite often Mr. Morris and Mr. Swenson met on the trains. They were great friends. Mr. Swenson has often reviewed his chats with Mr. Morris, pausing here and there to place special emphasis on some unusual bit of wisdom or practical thought of value to us. They were always making plans for visiting each other's properties, and I was to share in the treat; but something always happened to prevent the visits. I know how much Mr. Swenson regrets it, and to me it was a personal loss. Mr. Morris particularly wanted to see our system of calf winter maintenance in the days when we maintained in pens to sell and ship any month between November 1 and May 1. From the earliest day of knowing Mr. Morris until the present moment I have thought of him as one of the strong elements for good in the live stock industry. He was an inspiration to good, clean trading.

No American institution of commerce has had more wonders for me than the house of Swift & Co. While not detracting one iota from the splendid "carry on" work of his sons, until the house stands among the first commercial American achievements—and that means the world—everything can be reduced to one great personality, G. F. Swift. What is perhaps an equally great achievement is that of the perpetuation of a great business, handed down from father to son, in which we have three paramount illustrations in the packer world: the Swift boys, J. Ogden Armour and Edward Morris, who died and passed the work to Edward Morris, Jr.

Rich boys often enter into succession with the handicap of great accomplishments by their fathers. These young men go up through fewer opportunities to fight out great problems. Their environment is one of comparative luxury. They are plunged into responsibility. But all these men have shouldered their heritage and carried the flag of American commerce on and on, in spite of the criticism, baiting and often persecution which have come to them, doing as much as their progenitors in widening the avenues of trade and saving waste to the world.

The early story of G. F. Swift is well known. He was a peddler of meats in New England. Due to thrift and good business, he became a packer in Chicago in 1875, and extended his operations to Kansas City and Omaha in 1888—about the time that I went with the Kansas City Armours. His struggle for growth probably represents the most wonderful financing in earlier years ever done by an American firm. I shall not attempt to describe what I know about the details except to say that he was a heavy borrower from country banks, always met his obligations promptly, and had a perfect system of "kiting drafts." He finally arriving at his great system of thousands of stockholders—a credit as good as the best—and a business equal to his greatest competitor. I know G. F. Swift slightly, but his characteristics and methods, as they have come to me by observation and from many of his own men, have always

The late Philip D. Armour, founder of the house of Armour & Co.

fascinated me. They were basically sound and consistently followed. Once when I happened to be in Denver I dropped in on the manager of the Denver Swift branch house. He was an old Kansas City friend. In the course of our chat I asked if "G. F." had been around lately. (Packinghouse men almost invariably refer to the "higher-ups" by their initials only.) "I should say he had," replied my friend; "he caught me napping, and he sure landed on me. The bookkeeper went to lunch. I happened to think of something in the railroad office about half a block away, and wasn't gone over fifteen minutes; but when I came back I found 'G. F.' sitting at the desk, counting the number of pencils we had in use at the same time. He said: 'This must be a safe community if you can all go to lunch at once and leave the doors open.' Then, looking out at the wagons, which were all in by that hour, and lined for loading in the morning, he said: 'Is that just your idea of how Swift & Co. wagons should look?' I said, 'No, Mr. Swift; they need retouching, and I wrote Chicago last week for permission to have them painted.' Then I got my knock-out. He turned to me, and, with a little smile, said, 'Did you ask them if you could have them washed?' "

After all it is the little thing which makes the big ones, and I took the story back to the hotel that night, giving myself a good mental survey, and found that I had asked to have several wagons painted, but had overlooked having them washed. One of the most far-reaching works of the house of Swift & Co., done by the great mind of G. F. Swift, consisted in establishing the reputation for being "a good house to do business with." That meant high standards in products, fairness in adjusting claims, courtesy in every individual, or, putting it another way, to quote Marshall Field, "The customer is always right, even when he is wrong."

Mr. Swift's was the achievement of plodding persistence, gained by inches rather than by brilliant strokes, and the name of G. F. Swift will always be enrolled among the greatest captains of American industry.

The greatest individuality that I have ever come into contact with was that of Philip D. Armour. He was a partner in the Kansas City house, but the two houses were not consolidated until about the time I came to Texas. My association was purely a business one as against my intimate personal relation with his nephew Kirk B. Armour; but the two houses encouraged a close relation between the men in Chicago and those in Kansas City, resulting in frequent interchanging visits. Philip D. Armour always had the Kansas City men come into his office, and they usually left it with their conceit pretty well abstracted, and some very wise things to recollect. I know that it sounds harsh to read of the persistent policy of the Armours to "roast" their men, but as a matter of fact it was only the old Irish trick of whipping up the free horse. The men loved it because they knew it was a sort of pat on the back, and an admission that a fellow had something in him.

I have rarely known a more lovable man than Mr. Armour. He was the sort who loved those whom he chastened. His men worth while loved him, and realized that after they came out from a "once over," or even after getting the "third degree," they did so with new fighting power. They were nicely cleaned up, like men getting out of the barber's chair after a round-trip. No commercial eye in American trade history has had the vision, with the quick-action executive ability behind it, that Mr. Armour had. His was the kind of vision which has been behind all great trades. It was this vision which took him from the York State farm to the goldfields, and made a fortune in pork after the Civil War. This characteristic came under my own observation several times, notably before the panic, I think in 1893; but I am not where I can look up the year. For months before it struck he was accumulating gold in his Chicago vaults, and shipping gold to Kansas City. When the evil day came both the Chicago and Kansas City Armours bought live stock every day, paying for it in gold instead of the usual check clearance. Again in 1893 an attempt to break him was made. He had bought wheat heavily in the northwest. The sellers planned to deliver him the actual product largely in excess of Chicago wheat storage capacity. Reviews of this incident speak of it as the "crisis of his life." Architects told him that it would take six months to build the capacity needed. He brushed them aside, began work in a few hours, and in forty-two days was receiving wheat in the new elevator. Of the curing cellars in Kansas City old "Matt" Harris had charge. He was a wizard in his work, but could hardly read or write. He could figure, however. Often he was given some unusual problem to handle, and his invariable reply to K. B. Armour was "It is '*un*possible,' Mr. Armour, but we will do it." If I were to cast about for some phrase to illustrate the life of Philip D. Armour I should use "Matt's" words "It is '*un*possible,' but we will do it."

I recall a personal incident which I might run into pages, and leave it still uncovered. I was in the Armour branch house office in Los Angeles on one of my western trips, when "P. D.," as we spoke of him, came in with a party of Chicago friends. Among them was Rev. Frank W. Gunsaulus, with whom Mr. Armour had close association in his extended philanthropic work. There began a friendship which has been more than a joy, rather a refuge, to me ever since, and I think of him as America's greatest divine. I was asked to go with the Armour party to visit the Cudahy killing plant. In parting "P. D." said: "We are touring California and the north coast. Whenever you find our car on your train come

The late Nelson Morris, founder of the house of Morris & Co.

in and stay with us." About a week later I crawled onto a train at Sacramento at 5 a. m. I noticed his car on the train, but thought that I would wait until after breakfast before making a call. His personal attendant came through the train about 6:30 and recognized me. He returned shortly to say that Mr. Armour (always an early riser) wanted me to come back to breakfast and remain with them, as they were going to be run "special." I wish I might have space to describe that wonderful trip via Mt. Shasta, Rogue River Valley, and the Willamette Valley. The train consisted of an engine and the private observation car. Peach trees were in bloom, and spring's charm was upon the landscape and in the air.

After breakfast Mr. Armour took me into the back end of the car and asked, "Well, what have you been doing?" I was pretty well fixed; luck had been coming my way, and I replied that I had sold eighteen cars of hams the day before. "That's fine," he said; then came that vision in the remark: "You don't have to go any higher up than me for instructions, and I tell you to sell this stuff and never stop selling. We are going to be smothered with hogs this fall, and I want a clean house. You will of course confer with your house, but if you get in a tight place, and haven't time, sell; you know enough about value not to go wild."

I never knew what he wired my house, if anything; but the house evidently had his ideas in confirming my sales, and working on the basis of his instructions I had a lucky day in Portland, Ore., selling 500 tierces of lard, and closing the deal just ten minutes before I had the house's confirmation, as the buyer had put a time limit on his offer. He was right; we were smothered that fall, but we met the situation with our general products up to cure.

It was more than a year before I saw him again. I was called into his office for the "third degree." I believe that he would have made a good boxer, for as quickly as a fellow could get in a word in support of his position Mr. Armour side-stepped and landed in a new place, until one was over the ropes and dizzy. He began by saying, "I have been itching to get hold of you ever since we were on the coast. You were doing fairly well then, but you have gone plumb to the bad since." Then it came so fast that even Babe Ruth could not have hit one. When he had me on the mat, and I had taken the count, I said, "Mr. Armour, there is one thing I can't understand, and that is why a man of your wonderful business judgment should keep on paying wages to a fellow as rotten as I am."

He chuckled, shook himself, and said, "Why, that is just pure philanthropy," and then added, "Come on out and see the boys." He put his arm over my shoulder and, walking out into the old main office on La Salle Street, said, "Boys, I want you to be mighty good to Frank, because the 'old man' has been mighty mean to him." His face beamed with good nature, and I wondered whether, after all, I had not just dreamed that third degree, and yet I knew that a great mind had either wasted an hour on me or had given me some of its great wealth for my life's capital.

By a strange coincidence I knew Michael Cudahy better from a personal standpoint than any of the packers, apart from those in my own house, and yet I knew less of him from a business standpoint than of the others. I was very fond of him. He had a charm of manner, a gentleness and a polish which appealed to me. Many of my frequent chats with him were full of pleasant memories. I recall passing a part of an afternoon and an evening with him on the train. I was impressed by his comprehensive general knowledge, but I cannot recall a characteristic story. He was for years general superintendent in the Armour Chicago plant before my time. He was an astute judge and handler of men. Every contact I had with him left its impress of force of character, keenness of judgment and trade instinct. When he died I felt the keenness of a personal loss.

CHAPTER XVI

THE LONG TRAIL

IT IS not my thought to attempt to treat the trailing of cattle over long distances from Texas to Montana and the general northwest, or from Texas to Kansas, from a personal standpoint, because these movements preceded my advent into the range industry. My memories are of association with men who were active in that epoch, and in turn have given me their own backward vision, or I have obtained it from a careful study of what has been published in random reviews. For those who wish the story in detail I recommend a book just published by the trail drivers of Texas. It may be had of Geo. W. Saunders, San Antonio, Tex., president of the Old Trail Drivers' Association, at a price which he will quote on request. Two books written by Andy Adams, an old trail boss, furnish in fascinating detail all the incidents of a dozen trails crowded into a composite story of one trail. One of his books is *The Log of a Cowboy*. It tells of trailing from Texas to the northwest, while *Wells Bros.*, his other book, contains the story of trails to Kansas, and of the men who built up a business along the line buying sore-footed cattle. These volumes may be had through general booksellers. I have had several letters asking me if I can locate copies of the out-of-print McCoy book, to which I have referred previously in this series, and two men have written that they were coming to Texas to see the volume!

One of my fellow-townsmen made five trips on the trail, and many cowmen still in the range country made at least one trip while they were youngsters. There is rarely a convention of the cattle raisers of Texas which does not bring forward a series of reminiscent talks from some old-time trail boss. These talks recall that the freemasonry of the industry stood in those days for considerable latitude. I remember a story which illustrates it. A drove of 2,000 head was started on the trail from south Texas to Montana. The drivers arrived with 2,000 head, but about half of them were then in other brands than those with which they had started. The drivers had lost cattle in stampedes and had gathered their number, regardless of brands, as they went along. That was their code, but it must be remembered that beef had so small a value that, as Saunders records in the old trail drivers' book, "to be had for the asking." It was, too, all of the same general quality. "The Longhorn was in the heyday of his glory," and was practically on a par with wild game. I remember Ike Pryor's telling in one of his reminiscent speeches of an immense drove of aged cattle, bought at $6 per head on the Rio Grande, on which he lost money. The freemasonry then had its logic, and the scramble for the maverick was more of a game than anything else.

The same basic motive which started Texas herds north in 1868, or even earlier, applies today, in the fact that cattle taken from the south to the north make wonderful gains. A fat cow which will weigh 1,000 pounds on grass in Texas today will at the same age show from 200 to 300 pounds more on the bunch grass of Montana, perhaps not so much in Kansas or on cornbelt bluegrass, but certainly an appreciable increase. That is why the Texas feeder of today is popular. It is simply its instinct to put on heavy gains when taken north. I have heard cowboys tell of their horses which were sold at the end of the Montana trail. On returning the next year it was difficult for them to recognize even pet horses, which appeared to them to have grown taller and spread out, giving them the appearance of entirely different animals. Upon a casual glance at a Texas "remuda" (the range name for a bunch of cow ponies) after a year on Montana bunch grass, the boys said that they would not have recognized them as the same lot brought up the preceding season.

One of the difficult things that I have encountered in this chapter is an attempt to divide into epochs the trail industry to Kansas and the Kansas City market, and that to the northwest. "Joe" McCoy in his book published in 1874 records a small movement to the northwest, while scattered through the individual sketches of the old trail drivers' book I find records of trails to Cheyenne and even Utah in 1870 or earlier. After a good deal of reading on the subject, I think it fair to divide the trail industry into two sections: First to Kansas and Nebraska (1866 to 1880) and, second, to the northwest (1874 to 1889). There were undoubtedly interlocking trails and trails subsequent to 1888. Ogallala, Neb., perhaps comes in for the widest range of dates. It is spoken of in the data which I have been studying as "the northern market." I find in individual cases many records of cattle taken on from Ogallala to northwestern points, but usually sold there, and different outfits taking them on. In a general way, I think all trails to Ogallala must be thought of as reaching a point for distribution. In compiling what follows I have gone carefully over more than 1,000 pages of various early histories, mostly made up of individual experiences. It must be remembered that Indian troubles occurred more or less all the time, and that they were naturally worse on the northwest trail. Many trail bosses record keeping peace with the Indians by giving them a beef every day. I have not, however, attempted to portray the thrilling incidents of trails over eighteen years. Throughout my study of records, and in the course of talks with old trail drivers, dates have been the most difficult to obtain. Alvin H. Sanders in his chapter entitled "The Long Trail" in *The Story of the Herefords* evidently encountered the same diffculty, because he used no dates, and Andy Adams devotes himself to incidents, without giving dates. With this handicap I must beg the indulgence of my readers if apparent or real inaccuracies should occur, because it has been a tedious and time-consuming task to present the result of my research.

Jerry M. Nance in his reminiscences records that the trail to the northwest was drawing to a close in 1889. In random data I find the comment that the trail to New Mexico was closed in 1895, after having been used twenty-seven years, but from other data I find that the trail to the northwest diminished gradually from 1885 to 1889. I was in Colorado several times from 1880 to 1895, and had occasion to notice the rapid encroachments of fences during that period.

Texas was fencing rapidly; herds could of course trail through big pastures, but the old open range days were passing by leaps and bounds. The S. M. S. people built their first wire fences in 1882, and were among the pioneers. Saunders, in his introductory chapter entitled "Old Trail Drivers," says:

> In 1867 and for some subsequent years there were no wire fence or material enclosures from the Gulf of Mexico to Kansas; grass was knee-high, and beef was to be had for the asking. In Texas there was no demand for the longhorn or his hide, but in other states, where the population was

greater, both were needed. Trail drives were through regions infested with hostile Indians, who many times carried off the scalps of cowboys.

I find that the great preponderance of trail drivers, particularly the owners—trail bosses and straw-bosses—originated in south Texas. This was quite natural, however, because it is the oldest part of the state, from a cattle industry standpoint. The great bulk of the drivers seem to have made their homes at Lockhart, in Caldwell Co., Tex., about 70 miles northeast of San Antonio. It is estimated by Saunders that in twenty-eight years, beginning with 1866, an average of 350,000 cattle per year went up the trail from Texas or 9,800,000 all told, at an average of $10 per head at home; and 1,000,000 horses at an average of $10 per head at home, or, in money, $1,008,000,000 in twenty-eight years. Had these old trail drivers not sought out the northern market, cattle must have died at home, and the range been so overstocked as to cause much of it to become worthless. Too generous a tribute therefore cannot be paid to the few fearless men who took up the trail vigorously in 1867. Saunders concludes his thought by showing the influence on its followers in reaching high places by adding:

The ranks of the old-time trail drivers are getting thinner. They are scattered from Texas to the Canadian border, and from California to New York. Some are rated in Dun's and Bradstreet's in the seven-figure column; many are at the head of Texas banks or owners of goodly herds of well-bred cattle, while some are the result of the human average and ending their days in humble circumstances.

The Old Trail Drivers' Association now has its annual meeting at the same time and place as the Cattle Raisers' Association of Texas. It is a joy to see them together, and if one can get into the sacred circle where a knot of them is living stirring events over again in reminiscent exchanges, it beats all the fiction of the day.

To Saunders' comment about the overstocking of the range without the trail industry should be added a line on the American buffalo, which wintered largely in Texas. Chas. A. Siringo in his book *A Lone Star Cowboy* writes of having visited Amarillo in 1912, when a buffalo bull bought from Capt. Chas. Goodnight was hung up to sell at $1 per pound on New Year's eve. The author reflects that in 1877 he saw near Amarillo a herd that was estimated to comprise 1,000,000 buffaloes in one black mass.

At this point I must digress for a moment. Saunders comments that at the close of the Civil War the soldiers came back home "broke." Old men, boys and Negroes had taken care of the cattle, many of which were unbranded; the range was overstocked, but there was no market. I find, however, in other records that two cattle-killing plants were located at Lockhart, Tex., in 1861, and it may be of some interest to interpolate some personal Swenson history.

The elder Swenson was a pioneer Texas country merchant. He first located at Richmond, Tex., on the Brazos River, about 40 miles above its mouth in Fort Bend county, and by a strange coincidence his sons, the owners of the S. M. S. Ranch, now manage a sulphur syndicate at Freeport at the mouth of the Brazos, acquired in the last ten years. This property, with the Louisiana sulphur mines, furnished all the sulphur for the Allies during the World War, the Freeport company furnishing between 1,500 and 2,500 tons daily. The elder Swenson moved to Austin, and before the war to New York, where he started a bank, and also functioned as a sort of clearing house for Texas products shipped by boat, such as hides, tallow and barreled beef. The Swensons were also agents for all the cotton ties used in America, and for years carried 90 per cent of the Texas banks' eastern exchanges. It will be seen, therefore, that Texas beef and by-products were being marketed to some extent many years before the advent of the Kansas City packing houses.

I find records of venturesome spirits that took trail herds into Louisiana and other Confederate states during the war for use in the Confederate Army, and sold them at profitable prices, but were paid in Confederate money, most of which eventually was worthless, although it was current for the purchase of general products for a time.

J. N. Boyles in 1866 drove a herd from Texas to

central Iowa for Monroe Choate and Borroum. M. A. Withers drove a herd in 1859 from Lockhart, Tex., to Fredericksburg, Tex., and in 1862 to Shreveport, La., selling the cattle at $20 per head, delivered. In 1870 W. E. Cureton took 1,500 head to a point near Los Angeles, Cal., wintered them there, and took them on to Reno, Nev., in 1871. E. A. Roebuck went with a herd to North Dakota in 1876, and made a trip soon after to Utah. L. B. Anderson records that he was eighteen years on the trail, without exact range as to dates. A. N. Eustace made eight trips (1879 to 1887), comprehending both Kansas and the northwest. W. B. Hardeman refers to a trail to the northwest in 1886 with Blocker, Davis and Driscoll. The stock comprised 40,000 cattle and 1,400 horses. R. McCoy mentions 260,000 cattle as crossing Red River in 1866. He speaks of 35,000 cattle reaching Abilene, Kans., in 1867 or 1 per cent of the then estimated cattle in the United States, as calculated on the basis of probably meager Governmental data. Siringo records that in 1880 the "Old Chisholm" trail was impassable for large herds, as with plows the "fool hoe men" were turning its packed surface into ribbons. He also states that in July of that year he reached Tascosa, Tex., with a trail herd of 3,700 head for the northwest, and that during their rest at Tascosa the first cowboy, Cape Willingham, was killed, the incident forming the basis for the afterwards famous "Boot Hill Cemetery," in which only men who died with their boots on were buried. I wish I might pause to give more of its history, gleaned from printed facts and word-of-mouth stories told by old-timers.

In 1882 Jack Potter took a herd to Little Big Horn in Wyoming. R. D. Hunter in 1867 drove 1,200 head from Texas to Omaha, and sold the stock to Government contractors at a good profit; he also sold 2,500 head in Chicago in 1869 at a profit. In 1879 S. D. Houston took a herd to Ogallala, Neb., as also did "Jeff." D. Harris in 1881, which, in a way, defines the movement into the Dakotas through Ogallala as a northern market. I could fill pages with dates falling into my arbitrary epochs, but I have only sought to show their individual variation.

In 1873 the panic almost wiped trailmen to Kansas off the map, but in 1874, while the drives were lighter, the year was quite generally profitable. 1871 was a bad year, while 1872 proved successful. McCoy says that 450,000 cattle entered Kansas in 1873, besides 50,000 turned eastward at Coffeyville, Kans., for a Missouri Pacific Railway connection. The Kansas trail is confusing to some extent, as herds went via Baxter Springs, Kans., thence into southern Missouri in 1867 to 1869 to make a connection with the Missouri Pacific. In the same years trail herds went from southern Texas into New Orleans. These were in the main small, and the great mass went to Abilene, Kans., or to some other Union Pacific point, until the Santa Fe began to take them at Dodge City, Kans. Herds along the line were beginning to form, but from 1867 to well along in the '70's Kansas and Nebraska were the great objectives. During that period it was not so much a question of making money as finding a market for surplus stock at some price. Very little money was available for financing, so that buying on credit in Texas, the sums to be paid upon the return of the drivers, who gave no other security than their word and a list of brands, with the amount due, was the rule. There are many stories of the sharpers who operated at Abilene. Bogus checks were used, and cattle were bought to be paid for in Chicago. Every conceivable scheme known to the trickster was worked on credulous stockmen, whose word at home was better than a bond.

According to W. P. Anderson the Old Chisholm Trail was named after a half-breed called Jesse Chisholm, who ranched in the Indian Territory. In the early '60's he had driven a herd of cattle to the Government forts on the Arkansas River. The name has often been abbreviated and used as "Chism." Siringo in *A Lone Star Cowboy* gives the following as the origin of the Chisholm trail. It was furnished him by David M. Sutherland, Alamogordo, N. M.

In about 1867 the United States Government decided to move some 3,000 Indians (Wichitas, Caddos, Wakos, Anadarkos) to a new reservation in Indian Territory. Their camp was located on the Arkansas River, where Chisholm and Cow Skin Creeks flow into that stream. They were moved during the Civil War with Maj. Henry Shanklin in charge. He had made a deal with a rich half-breed "squaw man," Jesse Chisholm, to open a trail, and establish supply depots. Cow ponies were used to drive back and forth at crossings in treacherous quicksand streams, such as the Double Mountain and Salt Forks of the Brazos, Cimmaron and North and South Canadian, settling the sand and permitting the crossing of heavy wagons. [This process is still used in Texas, and during my early years in the state when driving across country we often unharnessed the horses when swollen rivers were encountered, riding them back and forth, and in later years I have taken off my shoes and tramped in the sand, still "quicky" after a rise, in order to make crossing in an auto safe.] With the advent of the trail driving the Chisholm trail was extended, and as nearly as exhaustive research can determine in the somewhat conflicting accounts by different authors it was as follows: C. H. Rust, San Angelo, Tex., says the trail started at San Antonio and ended at Abilene, Kans. The route was San Antonio to New Braunfels, thence to San Marcos, Austin, Right Round Rock, Georgetown, Salado and Belton to old Fort Graham near Waco; thence to Cleburne and Fort Worth, and on to Bolivar, where the trail forked out, most trails going up Elm to St. Joe on Red River; thence to Nation, Beaver Creek, Monument Rocks, Washita Crossing and Canadian River to the north Fork, Prairie Creek and King Fisher Creek; thence to Caldwell, Solomon and Abilene, Kans., crossing Cow Skin Creek and the Arkansas River and passing via Brookville.

The trail varied in width at river crossings from 50 to 100 yards to from 1 mile to 2 miles at the widest points. The average drive per day was from 8 to 12 miles, and the time on the trail was 60 to 90 days from points in Texas to Abilene, Kans. The other trail from Bolivar crossed the

Red River below Gainesville on to Fort Arbuckle, and intersected the main trail at the south fork of the Canadian. The last main western trail ran by Coleman, Tex., Bell Plain, Baird, Albany, Fort Griffin and Double Mountain Fork, crossing Red River at Doan's Store.

Saunders records another trail known as the McCoy trail, which started at Corpus Christi, Tex., and ran to Austin, Georgetown, Buchanan, Decatur and Red River Station, all in Texas, and thence to Abilene in Kansas.

McCoy relates that in 1870 1,400 selected beeves sold in Chicago at 4½ to 6¼ cents, netting $20 per head to the producer, and that in 1867 a buyer bought from a herd of 3,500 head his choice at $6 for 600 head, and a second cut, his choice, of 600 head at $3 per head, or an average of $4.50 for 1,125-pound beeves or 40 cents per 100 pounds. They were probably fours, fives and sixes. This is the only reference to weights during the early period found so far in my reading.

Photo by Reeves

S. M. S. Boys Saddling in Corral

CHAPTER XVII

TEXAS COW PONIES AND STUD HORSE LUCK

A LARGE number of Texas cow outfits have had their notions about cow-pony crosses, the word "pony" being applied to all sizes of horses, and many breeders have succeeded in producing some good ones; but, in the main, cow horse breeding has been a pure case of "scrambled eggs." It would seem to be hopeless to try to classify or try to get anything definite or rational in the way of distinct lines of breeding beyond the basis which can usually be traced to pure Spanish origin in Texas breeding. I am told that Colorado and the northwestern states trace their cow horses to the Mustang. Texas draws many cow horses from Colorado, but they are not liked so well as the Texas-bred horse by Texas cow people. My own investigations deal with Texas-bred horses.

Many ranch outfits do not breed horses, but depend upon purchases, as needed, usually picking up odd young horses, here and there, as anything suitable is found. In the main, however, they buy bunches from a trader or commission some one to put up a band under certain specifications. These outfits do not often have so satisfactory a ramuda as outfits raising their own horses.

I often think that old Steel Dust must turn over in his grave when his blood is made responsible for some of the broom-tails tacked onto him. It is still a name to conjure with, but almost every horse trader who has not recently joined the church or been rescued from backsliding will declare that his line of equine stock is largely Steel Dust in its breeding. There have been some outstanding Steel Dust stallions used in Texas, but for the most that blood has gone into cow bunch outfits through grade stallions.

From one standpoint it seems incredible that with so much depending upon the cow pony the whole plan of producing him should apparently have been badly neglected; and yet until ten years ago his value lay entirely in having him educated, or developed into a cow horse.

I recall that when I came to Texas in 1902 the S. M. S. outfit sold 200 mares at $7.50 per head, giving the buyer his pick; but it was really more a surplus than a cull sale. Value, therefore, in the early days had its influence upon neglect.

Talking with men in outfits that bred their own horses, I am convinced that the whole problem of breeding cow horses has been saved by "the law of selection" in sires—not registered or purebred stallions, the early-day sires being selected by men who knew a good horse, and used him because he was a good horse. I know of only one Texas ramuda which can be called a "type"—the 6666 outfit, owned by

S. B. Burnett of Forth Worth. It is the result of a carefully thought-out and followed line of breeding. Mr. Burnett and the Indian chief Quanah Parker formed a close friendship in early days, which lasted until that Indian's death not long ago. Mr. Burnett had something like an Indian's love for a "paint" horse, as he is known on the range, or pinto or calico or circus horse, as most know him. Quanah Parker furnished the basis, and Mr. Burnett has lived to gratify his dearest wish—"To drive a purebred herd of cattle to market with a paint ramuda;" but the paint ramuda is a story in itself, which I may tell as such at another time.

Every other Texas ramuda that I know contains all colors and classes: Horses that will weigh 1,100 pounds, others that look like schoolboy's ponies, and some of the best horses in the ramuda in both classes.

A brief history of the S. M. S. cow ponies will probably typify the haphazard methods of cow pony breeding. Perhaps some cowmen and cowboys of other outfits may smile when I classify the S. M. S. ramuda as "good." It has more size than the average ramuda, but is, I think, regarded as above the average, from a cow horse standpoint. In 1882, when the S. M. S. Ranch was started, a horse trader came through the country with what was called a good bunch of Spanish mares. Fifty of these were bought to start the S. M. S. breeding band. Spanish horses, as I understand them, were a pure Mexican breed—small, mean, tough, quick as a cat, and had the "cow instinct," which suggests that it may be well to say what a cow pony should be. His first qualifications are speed and endurance. Nevertheless, one-may have these and still not have much of a cow horse. He may be all right for ordinary rounding or line work, but he is not a cow horse unless he have "cow sense" as a dominant characteristic. Training has much to do with it, but he must have the instinct for holding a roped animal, "turning on a half-dollar," and countering every move of an animal that is being cut out. The old Spanish horses had this instinct as true as the bird dog has the bird instinct, and that is why Spanish blood is the basis of most Texas ramudas. Practically no producer of cow horses, however, appears to have been satisfied to stay with the Spanish blood, in its purity. The difficulty of getting satisfactory Spanish sires may largely explain this fact, but probably meanness had much to do with the popular desire to breed the strain up without losing the cow instinct.

To go back to the fifty Spanish mares bought to found the S. M. S. breeding bunch, a sire was needed, and as the early employees of the ranch had come from Williamson county they went back home for a stallion. I have their statement that he was Arabian. Certain it is that he possessed nerve, endurance, style and action, was "a horse all over," and pure white. Many of his get were still in use when I came to the

S. M. S. Ranch twenty years later, and from them I formed my early ideals of great cow ponies.

The Arabian sire, regardless of what he really was, illustrates what has probably saved the cow pony—"the law of selection." Some man with good horse sense had picked the right sort of sire.

For years the S. M. S. Ranch ramuda could be identified at a great distance by the predominance of white horses in it, and even today, when some special occasion demands, the entire outfit will come out mounted on white horses.

Several Missouri saddle horses were crossed on the daughters of Arab; also a high-strung Standard-bred horse. A good Thoroughbred followed. In 1901 pure Spanish-bred stallions from the King Ranch were put in, followed by some grade Percheron and Clydesdale strains from native mares. Many of the best horses in the ramuda are from these latter stallions, crossed on mares which represented Spanish, Standard-bred, Arabian, Thoroughbred and Missouri saddle stock, mixed beyond all hope of accurate classification. Then followed more Missouri saddle stock and Standard-breds, and a race horse from a Standard-bred and pure Thoroughbred cross, which had a mark of 2:14 in his first year, but went lame and came to us at a bargain.

A German Coacher, bred from the halter to mares selected for their fitness to mate with him for drivers, for work horses, and some experiments with cow ponies in view, has produced some excellent horses in all classes; but his chief value from a cow standpoint, will be in the cross of some of the Morgan strains, or the race horse on his daughters.

During the past five years, nine registered Morgan horses from the Richard Sellman Farm, Rochelle, Tex., have been added. They weigh 1,000 to 1,150 pounds, naked.

A constant culling of mares occurs in all ages. No mares are used in any of the ranch work, but a grading of geldings, suitable for farm work for drivers and for cow horses, is made, which helps wonderfully in concentrating in the ramuda the best cow horses. Probably 60 per cent of the horses cut out for cow work make fair cow horses, 20 percent really good ones, and 10 percent "crackerjacks."

The S. M. S. Ranch horse total is 1,100 head for all purposes; 500 head are used in distinct cattle work, and 100 for farm, freighting or team work. There are about 275 breeding mares, the remaining 225 being in various classes and ages coming on. The capacity of one jack is used for mule production.

The cow horse contingent takes care of 400,000 acres of pasture, conservatively stocked with cattle. Cow horses are broken as threes in the spring. Their average use in cow work is 12 years, and it is no exception to find horses 18 to 20 years old still doing good work, notably cutting horses, saved for that work, and not used for general riding. These fig-

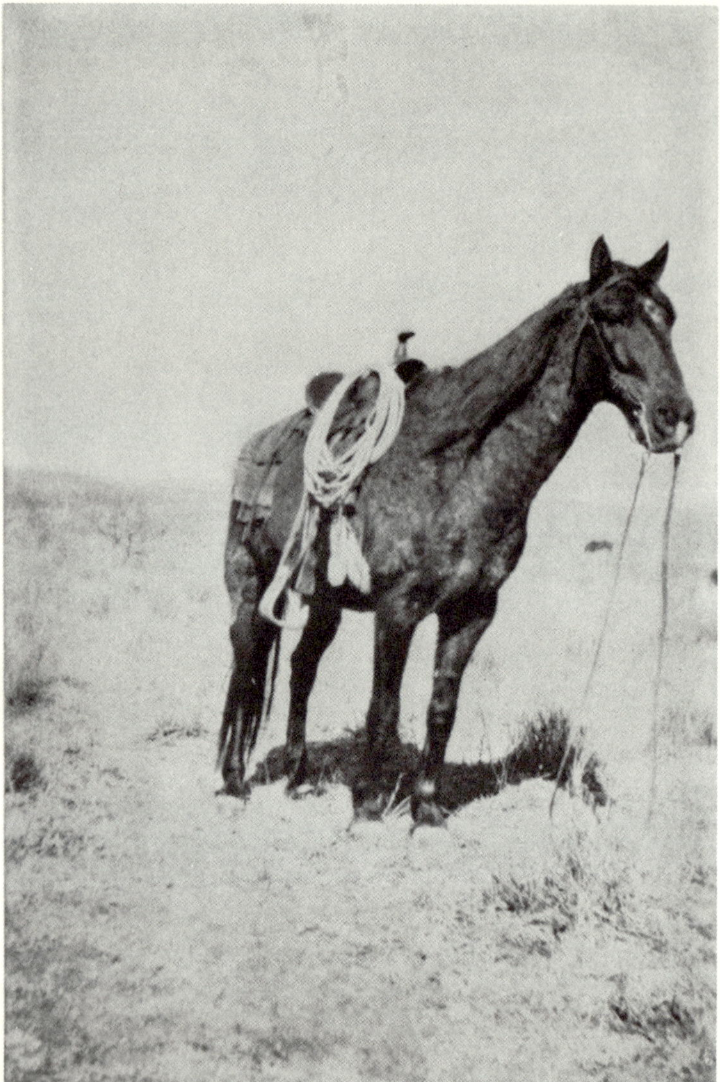

Photo by Vinson

"Ol' Cabby"—22 years old and still in service on S. M. S. Ranch
(See poem opposite page)

THE OL' COW HOSS

When it comes to saddle hosses, there's a difference in steeds;
There is fancy-gaited critters that will suit some feller's needs,
There is nags high-bred an' tony with a smooth and shiny skin,
That will capture all the races that you want to run 'em in.
But for one that never tires; one that's faithful, tried an' true;
One that allus is a "stayer" when you want to slam him
through,
There is but one breed o' critters that I ever come across,
That will allus stand the racket—'tis the Ol' Cow Hoss!

No, he ain't so much fer beauty, fer he's scrubby an' he's
rough,
An' his temper's sort o' sassy, but you bet he's good enough!
Fer he'll take the trail o' mornings, be it up or be it down,
On the range a-huntin' cattle or a lopin' into town,
An' he'll leave the miles behind him an' he'll never sweat
a hair,
'Cuz he is a willing critter when he's goin' anywhere.
Oh, your thoroughbred at runnin' in a race may be the boss,
But fer all-day ridin' lemme have the Ol' Cow Hoss.

When my soul seeks peace an' quiet on the home ranch of
the blest,
Where no storms nor stampedes bother, an' the trails are
trails o' rest;
When my brand has been inspected an' pronounced to be
O. K.,
An' the boss has looked me over an' has told me I kin stay,
Oh, I'm hopin' when I'm lopin' off across that blessed range,
That I won't be in a saddle on a critter new an' strange;
But I'm prayin' ev'ry minnit that up there I'll ride across
That big heaven range o' glory on an Ol' Cow Hoss.

—A. A. Brininstool, in *Los Angeles Express.*

ures are based on horses which live, and do not comprise the fools or outlaws, which are cut out and sold in what is known as the "scalawag bunch."

At one time the demand for polo ponies, and the idea that Texas cutting horses filled the bill, threatened to make a serious drain on our cow horse stock, because the required qualifications took the best cow horses; but the situation was saved by the specification for height, and the fact that most outfits would not part with their best horses at any price.

The demand for war horses used up much of the good floating cowboy stock to be picked up young and untrained. It was usually raised by young country boys. If an outfit had to go out and buy a ramuda of 100 horses, the units would be difficult to pick up at any price in the country.

An outstanding cow horse is worth what he will bring, and it is inadvisable to quote a price which is thought too high for the other fellow to pay. I have seen more than one man who priced a horse too high back clean down, rather than let his horse go. There are more buyers for a $300 cow horse than there are horses good enough to realize that price or owners willing to sell.

I have been hearing all my life about "stud horse luck," but with never a thought of its origin. It comes home to me now as I think of S. M. S. luck in getting that first good horse Old Arab, and that leads to the conclusion that, after all, "stud horse luck" may mean that instinct has much to do with what we commonly call luck, and that the instinct in knowing a good horse may have been a definite plan of breeding in the production of many good Texas cow ponies by many outfits.

CHAPTER XVIII

THE STORM: A "MAGE" STORY

IT WAS late in October. We were going through the unusual experience of delivering late in the season to a Dakota buyer some 900 yearlings. He had made his cut, and driven off to look at some other stuff, while I started to town to see that everything was ready for the shipment next day. The outfit trailed with the herd to throw into a hold pasture at headquarters for the night. The afternoon had been sultry, and I noticed great thundercaps piling up in the northeast, but wind and weather do not count much in west Texas when one has something to do.

Summer had been rather on the yellow order, and the old Indian sign, "clouds all around and rain falling down in the middle," formed about the only weather bureau that we took much stock in. So I did not pay much attention to the kindly warnings as I passed headquarters. I rode the chestnut mare Beauty, a handsome, wilful beast, the fastest in the country for a long-distance run. She had her own ideas about who should run things, but she always responded to the call of duty, and her endurance was a wonder. We had made many a hard ride to see the baby go to bed, and the hour of daylight seemed ample in which to make 12 miles to town. I did not care to get caught in the mesquites after dark on a bad night. Things did look pretty bad, though. The atmosphere was sullen; scarcely any air stirred. There was a sort of oppressive hush. Greenish and yellowish tints hung like veils of vapor about the clouds. The dying day sent its flickering shadows like a ghastly smile, as if to say, "You and the night for it. I'm off." I gave Beauty the word, and we hit the trail at a strong distance-covering gait, which I knew she could keep.

A mile traveled, and she was sweating; the air seemed almost hot. Half a mile further we passed a settler's house which looked mighty good in the deepening gloom, but we had started, and the storm might go around. A quarter further and we shot into a cool current of air, as if it were out of a hot bath into a cold room. The mare shivered; then we met a distinctly cold breeze, and on came the storm. Beauty turned as it struck, but responded promptly to the rein, faced it for a few yards, and then reared and turned. Again she took it, this time with the bit in her teeth, and went at it as though trying to jerk a load through a mud bog, her blood up, and fighting hard. My head was almost on my breast; the rain beat with fierce force, and I was glad after ten yards of fight to have her turn again. We seemed to be in one glare of electricity and a roar as of a battle. We made for the settler's house. The horse lot was open and lightning struck one end of the house gallery as we dodged under the shed.

I waited for a lull, and rode back to headquarters. The boys were just getting in; the storm had struck them just after throwing the cattle into the hold pasture, and all they could do was to turn their backs and take it. During supper the storm came back from another direction; for an hour the winds blew; the heavens seemed one immense water bag, with the stopper out; lightning sizzed, thunder roared. We all sat around the fire drying our clothes. Most of the boys did not say anything, but one spoke of the big revival down at Fairview and "'lowed" that the outfit should all go down on Sunday. Another spoke of how much good Sam Jones had done in Texas, and that he was glad to see so many new churches in the county. Sam Sawyer told how he never had felt so good as when he helped pull a "nester" out of Double Mountain Fork when he got bogged down.

With occasional shots as from a retreating battery the storm passed. One of the boys asked another for some paper and tobacco; several took a chew, and "Abilene" remarked, "That was a pretty doggoned hard storm." Somehow a weight seemed to have been lifted, and a sort of "come alive" feeling was apparent as opposed to the "drop a nickel in the contribution box" look that everyone had been wearing for an hour.

Mage got up from a "hot roll," whereon he had been resting, and stretched himself, and when Mage stretches it means six feet five inches of all cowman grinding his bones.

"Fellers," he said, "as far as thet's consarn, I've allus noticed thet there's nothin' like a stampede in the skies to rope a cow camp down into a sort of revival meetin', an' I don't recall hearin' anything stronger than 'doggone' or 'dad gum,' come up during the millin' of a bad storm, 'cept onct, an' thet was the time we lost a herd on the Jane Wilson League."

Everyone settled down for a story. Few men in west Texas have so much material or can put personal reminiscence into so vivid and charming a form as Mage. The writer can only attempt vaguely to convey the theme; the sidelights of personality, intensity and humor, deep humanity and unconscious dramatic force can only be had when one hears Mage tell his own stories.

"As far as thet's consarn," he went on, "I've been thinkin' about it ever since this here storm commenced to round. I have never seed two days more alike, an' it was jest about this time o' year. We had a steer herd—threes and fours—takin' to Seymour to ship to K. C., about a thousin', all wild as a snake, an' some rank outlaws among 'em, too. I was only a kid then—the 'S. M. S. Kid' they called me, but I could ride some, and they took me along. But the outfit—thet was an outfit, all of 'em foremen now—there never was no better, an' them was the days of outfits, too. But when enyone tells you 'bout a foreman thet never lost a herd I'm a-goin' to show you a foreman thet never trailed much or throwed in with Providence on fair weather.

"We got an early start; the steers traveled off brisk; the outfit was feelin' good an' the 'cussy' put up a regular Thanksgivin' dinner. The mornin' was fine, but about two in the evenin' the breeze layed 'em down an' thundercaps begin to pile up like high pinnacles. The steers seemed to sniff somethin'; the air got thick. We watered at the lake on the edge of the Jane Wilson League, and give 'em a long, loose herd up to the bed ground, jest as the sun was makin' camp—and sech a sunset! There was somethin' creepin' about it; made you think of cities on fire an' big clouds o' smoke, with the red o' the flames showin' through, an' then it died into a sick-like yaller, with green trimmin's. The thunder was a moanin' and mutterin' like a sick hound off in the mesquites. There was a stiddy flashin', and now and then them forked boys would work out like a crazy man throwin' a rope. It was dead-still, like waitin' at a funeral for the preacher to begin' and everybody was quiet, too, 'cept Obe Hogan. If anybody ever starts a school to teach cussin' and swearin' Obe sure gits the high job. He kept commentin' an' remarkin' about things in general an' the clouds in particular in a way thet

I didn't take kindly to, hearin' an' lookin' at thet sky at the same time.

"But the storm didn't break like it did tonight; jest kept a hangin'. The foreman 'lowed we better catch our best night hosses and sleep without much undoin'. I ain't lost no cyclones, an' all the boys 'cept Obe was about the same mind; so we jest sits aroun' and drinks coffee an' eats tea cakes till the cook druv us off. The first guard seemed to be gittin' on all right, an' we fell asleep. 'Bout an hour 'fore midnight the foreman calls us an' says the cattle was gettin' restless, an' all hands went on duty. The air was still hot, and no stir. The lightnin' had about quit after sunset, 'cept flashin' some.

"We hadn't much more'n got to the herd when the air freshened an' things was gettin' right. Then it got cold, an' we could hear it comin'. Thunder and lightnin' seemed to spring out of the mesquites. The foreman passed the word: 'Hold 'em till they git wet,' an' we began to circle. The cattle was on their feet in a second, with the first cold air; but we got the mill started by the time the storm hit. I've seen lighnin', an' this little show tonight was a purty good imitation, but thet was lightnin' right. As far as thet's consarn', I've seen balls o' fire on the end of a steer's horn many a time, but there was a ball o' fire on the end of both horns on every one of them thousin' steers, an' the light in the balls of their eyes looked like two thousin' more. Talk about a monkey wrench fallin' from a windmill an' givin' you a sight o' the stars, or one of them Andy Jackson fireworks clubs puttin' off Roman candles at a Fort Worth parade! They're jest sensations; this here show I'm telling about was real experience. We seen things.

"Obe Hogan hit the herd a cussin', and was circlin' between me and Sam Conroy, an' the way he turned out new ideas in advanced profanity made my blood creep. I was sure if lightnin' was goin' to hit it would be purty close to where Obe was, so, bein' in front, I kept movin'; an' Sam, havin' about the same notions, tried to hole back, but somehow we kept bunchin', and Obe kept rippin' out an' gettin' better all the time.

"Hold 'em? Well, we tried. We circled with the drift, and when they broke we tried to slick 'em, but the storm kept getting worse an' when one of them hot-off-the-griddle boys come they was off an' we was flounderin' over unknown ground, trustin' to heaven more than enything else. It was either black as pitch or the lightnin' blinded you so you couldn't see. I rode old Blutcher, an' when one of them outlaw shocks would come he'd squat till my feet trailed. Three of us tried to head a big bunch an' I knew one of us was Obe. He hadn't got tongue-tied none during the storm; seemed like he'd caught his second wind. Onct for a while I didn't hear him, an' I sez to myself: 'Mage, it's lucky you didn't git hit, too.'

"Purty soon it was no use; they was gone. A mesquite limb brushed me off, an' when I lit Sam Conroy was squattin' on the other side. I squatted, too. Lightnin' hit a mesquite close by an' we moved out some, an' set takin' the storm for a long time without sayin' a word.

"A storm reminds me of an ole cow on the prod. She'll fight for a while, but she loads, an' the storm was gettin' ready to load. It quit howlin', and begin to groan, an' was passin'. Sam turns to me an' sez, 'Mage, I've been tellin' the Ole Marster that if He'll jest take me through this all right I'd try to make it up to Him in the way I lived from now on,' an' I sez, 'Sam, I'd a' liked to throwed in with you on thet,' an' Sam sez, 'Mage, I knowed you would from the way you was dodgin' Obe durin' his storm conversation, so I jest put your name in the pot.' An' I sez, 'Thank you, Sam.'

"The next day after we got the herd together, me an Sam was trailin' together, an' his hoss stepped into a badger hole, givin' Sam a fall, an' he lets out a few that wasn't jest parson talk, an' then he thinks an' sez, 'Mage, I'm a-goin' to make it up to Him all right, but I guess I'll have to hobble myself and put on a hackamore.' "

CHAPTER XIX

SOME HUMBLE WESTERN CHARACTERS

EVERY once in a while I pick up a literary review which takes a shot at Bret Harte. All reviewers credit him with the production of real literature, but some charge him with a gross exaggeration of his western types, which they characterize as unreal, and the creatures of a romantic mind. Then I know that the reviewer has probably looked out of a Pullman window on his way across the continent, been wined and dined in some of the western clubs, wherein he has probably met the very types that Bret Harte described, but did not recognize them in their dress clothes, or know of their earlier lives. Bret Harte wrote of the long ago, when one could squat in any mining camp, and find his characters. The west is probably becoming the east very rapidly, but it still has its "John Oakhursts," as real today as then, but harder to find, and the average book-reviewer could not get in touch with them.

It has been a felony for many years to gamble in Texas. I think it is so in most western states, but the west is still the west in most of the traits and characteristics which dominate the frontier. I recall an incident in Montana—just a little thing which illustrates what John Oakhurst did in a bigger way. He had the trait of sacrificing himself to protect others, if necessary; it was not courage but just a plain inherent willingness to do one's whole duty in a pinch, and take a chance. Perhaps it is not fair to call it western but rather American as our boys who "went across" have told the story in their deeds. Perhaps it is just because the west offers more hazards. I had gone with some friends over a high divide to an interior town near Missoula, Mont. Part of the road was along a canyon where one could look down a sheer thousand feet or more. The road in places was very narrow. There were cut-outs for passing. It was not bad in the daytime, but our business detained us, and nightfall overtook us before we reached the divide. There were four of us in an uncovered hack, a sort of spring wagon. The moon would not rise until late. I suppose that we were careless, laughing and talking, and telling stories, but we came to a sharp silence when our driver, an old mountaineer, stopped the team suddenly. We were well up on the divide, but some distance from a passing place. Out of the darkness came a voice, "Hello!" Our driver replied, "That you, Hank?" Then from the darkness came this: "Hello, Billy; unharness; lead your hosses by me; take off your wheels; throw your old cart up agin the side of the mountain, and *I'll go round you.*" All

Hank did was to come and look at the space with a lighted match or two and tell our driver to stand on our upturned wagon and whistle. Then, with the aid of a little starlight, he drove his four-horse team and freight wagon around us, death yawning a thousand feet below, and only inches to spare. It was a chance, but Hank had the guts to take it.

When I first saw "Joe" Bradley we were all lined up at the old Florence House Bar, drinking near-beer, or what was then perhaps nearer than now. I say we, but Joe took a cigar. He was a small man, both short and slight, and if one were looking for protection Joe would be one's last likely choice. We chatted together quite a while. He was mild-spoken and unassuming; there was a sort of gentleness about him, particularly in his eyes, which gave one the impression that he was just an every-day, modest good fellow. As we walked away, my friend said: "I am glad you met Joe. He is the 'nerviest' man in Montana. He runs a train into the Coeur D'Alene district, with a tough gang all along the run, and they all eat out of his hand."

A few days later I had occasion to go up into the Coeur D'Alene country on Joe's train, which made the round trip daily. It was through that wonderful St. Regis River country, wilder than nature itself, where the trout leap in the white waters, and hide in dark pools. As we came into the wilder parts the only stations were mining camps. There had been a strike in the mines; the men were visiting the different camps, in which "booze" flowed pretty freely. Between stations Joe and I had been sitting in a double seat, smoking and chatting. I faced the front of the car. He came into the front door after leaving a station, and was about half-way down the aisle taking his fares when a shot rang out a few seats in front of him. I happened to be looking at him. The change in him was incredible. He seemed to lengthen and broaden, and those gentle eyes had a gleam in them like that of a tiger cat. Joe moved so rapidly that he seemed to be at the seat where the shot was fired without moving. He was calm as death itself. His voice was steady and unraised, but there was a chill in it as he said, "What are you shooting at?" "Nothin'," came the reply; "just seeing if I could shoot through the roof and make you jump." "Ticket," said Joe. The shooter passed up to him the ace of spades. Again I didn't see Joe move, but the tough straightened out: the butt-end of a forty-four had stunned him. Then, backing himself up against the car, Joe said: "This is Joe Bradley's train; is there anyone else here wants to try to run it?"

I am not going to believe in mild eyes again. Joe's were searchlights, and by some sort of fascination every eye in the car looked into his, and there were no takers. Joe called for a cup of water, a little of which he dashed into the stunned man's face; then, as he came too, carefully washed the ugly wound. Joe's eyes had gone soft again, a little smile played about his mouth as he propped the hurt man up into a comfortable position, and said, "Partner, you must be a new-comer; one of those hellraisers from the Leadville trouble. This is Joe Bradley's train—'the Bradley special' for everybody who wants to behave hisself, and it's the Bradley funeral train for them that don't. Tickets!" and they came up promptly.

When we were chatting again I said, "Joe, how often do these things occur?" He smiled that mild-mannered smile again as the answer came, "Not often—just when trouble-makers come in; but that's the first one I ever really hurt; I usually just break an arm with the butt of my gun. I have never had to fire a shot."

Going across the continent thirty years ago, as I reached Spokane, friends began to ask if I had met Fred R. Reed. For a week or more on my way up the coast friends asked the same question. At Tacoma I got in about 10 p. m. As I walked up to the register in a hotel a tall, muscular fellow was chatting with the clerk. The register was slightly turned as I wrote, and the big fellow said, "For God's sake, are you Frank Hastings? All of our friends have been trying to get us together; I am Fred Reed." A friendship thereupon began which has lasted through the years. He was then working on a great irrigation scheme at Prosser, Wash., now a town of several thousand, but then just a wide place in the road. He was fifteen or twenty years before his time. The whole Yakima district is now a great wheatfield or fruit orchard. Reed was in every way an unusual man; a brilliant writer, a forceful speaker, a ready wit, and a charming companion. He had worked as commissariat in the railroad camps of the Northern Pacific during the building of the line. Later, when the Prosser bubble burst, he took to the mountains, and two stories that he told me when we last met some years ago are so typical of the west that I wish to make them a part of my sketch.

In all former frontier types I have touched upon men widely known. This chapter deals only with those known but little, and it is to the thought of what the frontier does in moulding types that I want to lead my readers. Zane Grey in a vivid story of the U. P. Trail has made a composite picture of early-day railroad camps, which in these days it is difficult to believe possible, and yet in all its lurid detail it was probably toned down to be printable. Fred Reed saw the drama during the building of the Northern Pacific, and I have seen it in the new mining camps, just as Bret Harte, saw it in the gold rushes of his day, and Jack London saw it in the Yukon. Fred called his story, "Our Funeral," and told it as follows, in his graphic way:

In addition to my camp duties I occupied the exalted job of justice of the peace. Now a J. P. in those days in Montana was a bigger man than the Chief Justice of the United States today. He had a perpetual variety entertainment. He married people, buried the dead, put out fires, took a drink with everybody, refereed dog, rooster and prize fights, and settled family rows. In fact, he did everything but stork work. He was called judge, but if he made a wrong decision his name was Dennis. One cold morning a bunch of gamblers waited on me, and said one of the girls in the red-light district had died in the night. Her's was a pitiful story. She was an educated, refined girl, who had married a scoundrel, who in turn deserted her in a mining camp, and the rest followed, even to the empty morphine bottle clasped in her cold, dead hand. I don't want to uphold the gambler, but you know the west, and that some of the best men in it are or have been straight gamblers. I have never been more touched than by the appeal of these men. One of them said, "Judge, you know it wasn't her fault; she just got her money on a dead card, and a crooked dealer took it down. Help us give her a funeral just like what she'd have at home, and, judge, we want you to say a prayer."

I shall never forget that gathering. The clean and the unclean, the bad man and the good man all stood together by that little grave in the frozen ground, and seemed to huddle together as by some human impulse, caste forgotten. A single white geranium had been pinned upon her breast by some Christ-loving woman, and it seemed to say to all, "Her sins shall be made white as snow." I choked on the first few words, and then it seemed as though a voice in my heart was speaking, and I heard it say: "Oh, God, here, amid the rocks and the pines, with the awful stillness of the mountains until the last day, we are laying away all that is mortal of a poor, tired little girl. Her soul is on its way to kneel at mercy's feet. Take her, O Lord, into Your arms; wash away the scars and bruises of a child more sinned against than sinning. Teach us Thy mercy; forgive her, and help us, for Jesus' sake."

Through my own blurred eyes I could see that they were all weeping, and one by one they dropped on their knees to the last man in silent prayer; then, without a word, they filed away. It was a quiet night in camp; every place of business, every saloon, every gambling house, every brothel being closed.

Another Reed story illustrates so vividly the loneliness of the mountains in which many men have lived alone, or with a dog, that I am sure it is worth reproduction. During one of our meetings I said, "Fred, don't you get awful lonely sometimes?" "Lonely?" he replied, "I avoid it all I can, though sometimes it has to be done; but the most terrible experience that I ever had was after my dog Prunes went out. I was camping near the Snake River in Idaho, in charge of a bunch of cattle. My camp was remote, and the only companion I had was a cayuse dog, an ugly brindle, white-eye cur; but he was loyal and true blue. I loved him and he loved me. I called him Prunes, because once, when I went for chuck and left him to guard camp, I was detained by a storm, and he ate all the prunes. He had his choice between salt and prunes, and he took prunes. Anyone who has lived alone, miles from a settlement, will understand why I loved Prunes. He slept with me, ate with me, and for five months was my sole chum and confidential friend. I told him all my hopes and fears, my victories and failures; he would grin and wag his tail as much as to say, 'You are all right, and I believe in you.' Grub was getting scarce, that is, fresh meat, and I decided with Prunes at my heels to go out and kill a jack rabbit. We had not gone more than 200 yards from the house when I got my first rabbit, and then, walking along rapidly, as dusk was coming on, I saw what I took to be another rabbit about forty feet away, and fired. I heard a howl of pain and anguish, and Prunes was done for, and I was alone. I lifted him in my arms, carried him to the cabin, laid him down, sat down beside him, and cried. How still it was! I have been lonely before and since, but that night was the longest, most lonesome and the dreariest that I have ever spent. I dug a grave by the doorstep the next morning, but I just couldn't stay. I loved him, he loved me, and trusted me, and I had killed him. I have been offered lots of fine dogs since, but have never accepted one. I guess my heart is out there in the sagebrush with Prunes."

CHAPTER XX

"GRUBB, THE POTATO KING"

IN MY home we are never quite so happy as on the days when we have baked potatoes, done to a mealy nicety, broken open, treated with a great chunk of butter, and served piping hot. Then my wife says, "That's all we are going to have today except pie, but there are plenty of pies." On those days we always add to our simple little grace, "And O Lord, please bless Eugene Grubb." I suppose that most people think of Mr. Grubb immediately as "the potato king"—the man who knows more about potatoes and has done more for potatoes than any one else in the world. When a man comes to be the best authority in the world on anything, from picking a time lock to saving a human soul, it means that he has had lots of competition, done a lot of thinking and hard work, and has not been in the vacuum class as to brain. Most men, however, are specialists. Perhaps Pope was right, as to the average, in his lines—

> One science only will one Genius fit,
> So wide is art, so narrow human wit.

But Pope missed it as to the great big-brained men whose lives have been broadened out into many avenues of achievement, and he missed Eugene H. Grubb by a wide margin.

Elbert Hubbard, who went down in the ill-fated *Lusitania,* left from his gifted pen a sketch of Eugene Grubb which is among the best biographical classics of recent years. In it he reviews the crucible in which the Grubb character was formed, the forge in which it was hammered out from the humble Pennsylvania Dutch farm to the meeting of Kings and Emperors, while making a report for the United States Department of Agriculture upon the comparative farming of Europe and America. The undercurrent of the work in blacksmith shops, on canals, rivers, derricks and farming was this: "This thing can be done, and must be well done."

It is my purpose in this sketch to go into the more intimate and human side of this outstanding American. I first met Mr. Grubb at the International in Chicago in 1904. I had known of him and his work, of his Colorado Shorthorns, and of his constructive methods in everything that he undertook for years; but this was my first personal contact with him. He had been showing an unusual load of two-year-old feeder Shorthorn steers at the St. Louis World's Fair, winning the grand championship at a time when there was real competition in the feeder classes at all the great cattle shows. I think that the highest compliment that I can pay his load of steers is to say that they looked like the sort of steers that Grubb would produce; they were "dandies." We were showing S. M. S. feeder calves. Since it was in the days when we reserved the right to pick a show load, we had gone the limit, and knew that they were good. I took one of our men with me over the show pens to look at what we had to go against in the sweepstakes if we won in our class. When we reached the Grubb steers I said, "Here is where we die." A man stepped up to me and said, "I think that we know each other; I am 'Gene' Grubb."

Personalities are dominant; this was a personality. He was tall and well-made, with "the head and front of Jove himself," a face beaming with that quality which made "Ben Adhem's name lead all the rest," the love of his fellow-men was written like a scroll upon his countenance, justice and fairness flowed from eyes set wide below a high forehead, and the square jaw told that this man would go to the last ditch. His voice had the quality of magnetism; it was soft and clear, and full of friendliness as he invited us in to "feel" the steers.

From his pens we went to our own, and he did me the compliment to talk more with my companion, who, he realized, had been vital in handling them, than he talked with me. So began a friendship which has stood the test of time, and brought to me, written on last Christmas day from his den in Butte City, Calif., where he now spends most of his time, a beautiful letter, graciously reviewing our first meeting, and the turning of the tide to young feeders.

That was an historic event. John G. Imboden was the judge. He gave us first in our district over some classy L. S. calves. He gave "Gene" Grubb first in his district on the twos (Shorthorns), and gave us the grand championship. It will be recalled that in the fat classes only threes (steers) had a look-in for the grand championship in the early days of the International; then it fell to twos, and today only yearlings seem to have much of a chance. I stood by our pens almost stunned to have beaten the magnificent Grubb two-year-olds, and looked up to see their owner rapidly approaching. His face beamed; he took my hand in a regular Methodist squeeze, slapped me on the back with the other, and said, "With all my heart, old man, I congratulate you. It was a just decision and a deserved victory." There was not a trace of disappointment in his attitude. To all his friends he said, "They won on their merits. If I had been judging the cattle myself I should have placed the ribbon the same way."

I do not think that John Imboden has ever been in a harder place. He had two perfect loads in different classes to pass on, and I have always felt that his decision was made on his early recognition of the drift to younger cattle. But what of Grubb, who lost in what seemed a foregone conclusion? A good loser, a sport like that, is real, and he writes his own character in letters that "he who runs may read."

Eric Swenson, my beloved young friend, who died in Colorado Springs last February, visited Mr. Grubb at Mt. Sopris many years ago. A story of that visit is another sidelight on the simple traits of fairness and humanity in the lives of these worth-while men. They had been knocking about over the country, and stopped at a little town. A crowd had gathered about a blacksmith shop, and a justice of the peace was trying for his life a dog suspected of sheep-killing. Eric and "Gene" were attracted, and found that it was a case of strong circumstantial evidence, and that the dog had neglected to retain an attorney; nor had the judge appointed any one to defend him, and things were going bad. "Gene" asked to be permitted to defend the dog. "Gene" does not know it himself, but those who have heard him lecture or speak know that when he unloosens himself he is eloquent, and Eric said that "he sure unloosened" on this occasion. His plea for humanity, his tribute to the dog in the abstract as man's companion and friend, rivaled the late Senator George G. Vest's classic. He carried the house by storm, obtained another chance for the dog, had a guardian appointed for him, and helped wash up the crowd, which had melted into tears.

Is it not beautiful that really great men are simple men, with the fear of God and the love of their fellow-men and dogs in their hearts? Which would you prefer, to have your name go down as the owner of great riches, or to be remembered like that of

Thoreau, or John Burroughs or Eugene Grubb, when his day shall come? Somehow the memories of those who have worked in the great outdoors will carry the perfume of the flowers, the song of birds, the low whispering of the breeze, the soft patter of the rain, the rustle of the trees and the murmur of the brook longer than those who have grown great and given much to the world from within closed walls.

CHAPTER XXI

REFRIGERATION AND BY-PRODUCTS

I HAVE received many interesting letters containing extended information on several subjects that I have touched. I had intended to devote a chapter to a sort of jack-pot condensation of this correspondence, but the book is drawing to a close, and my ranch work is coming so fast that I shall only take space for refrigeration. In an early chapter I promised to try to get data as to refrigeration at Denison, Tex., long before it seems to have been used by what are now the "great packers." Data have come to me through Warren V. Galbreath, general live stock agent of the Missouri, Kansas & Texas Railroad. He began his cattle experience as foreman of feeding stations thirty years ago. I have quoted from the best data that Kingan & Co., Indianapolis, in 1885 used the first mechanical refrigeration, and, so far as I know, that stands as applied to the use of cold air pumped through a plant. The use in packing houses of mechanical ice, however, as a substitute for natural ice antedates the cold air process by probably twelve years, and apparently found its earliest use in the south. Its first application to the packing industry seems to have been at Denison, Tex.

It is especially interesting to quote P. H. Tobin, manager of the Crystal Ice Co., Denison, Tex., concerning what appears to have been a well-defined attempt to start a packinghouse near the source of supply in 1874. With him I was put in touch by Warren V. Galbreath. From several letters received from Mr. Tobin I quote:

In the early '70s—1874 or 1875—there was a packing-house at Denison; there was also a number of refrigerator cars, twenty-two, I think, numbered in the 6400's of the "Katy" (the Missouri, Kansas & Texas) railroad. They were all painted white, and had an icebox run in the middle on the top of the cars. The railroad boys feared that the seven crews running out of here would be reduced to one run a week, as we were putting from 16 to 20 slim cattle, all wearing horns, in a car, and the slaughtered product would make a big reduction. The business failed because there were no re-icing stations until Kansas City or St. Louis was reached. The packing business at Denison was started by T. L. Rankin and a Mr. Bushnell, who were supplying natural ice out of Hannibal, Mo. Mr. Rankin put in a little four-ton ice machine in 1876 in the Compress Building, and I was leased with the building. We sold ice at 5 cents a pound at the plant, and went "broke." Mr. Rankin was persistent, and came back to Denison in 1880 to erect an ice plant.

Mr. Tobin also sends me a letter, written to him by R. S. Legate, president of the National Bank of Denison, from which I quote:

I came to Denison in the fall of 1874. The Atlantic and Texas Refrigerator Car Co. was operating a slaughtering plant here, and, for a brief period, seemed to be doing an extensive business. I do not remember how long it continued, but my recollection is that it was short-lived, and the "big idea" upon which it was founded did not materialize.

From the foregoing it will be seen that refrigeration in natural ice or mechanical form is the underlying principle upon which the whole packinghouse edifice has been built, and the failure to have a few icing stations probably cost Texas the long wait until Fort Worth came into the business. Who knows what it may have meant to Denison and the cattle industry?

C. E. Clapp of the Pennsylvania Railroad writes from Chicago as follows:

I believe it is generally conceded that the first attempt to transport perishable property under refrigeration was in 1865, or ten years before the date given in your article. In Andreas' *History of Chicago,* Vol. III, page 602, the story is briefly told in the following language: "William W. Chandler, general agent of the Star Union Line (the pioneer of through freight business as now carried on), came to Chicago in June, 1864, to assume the position he still holds. He came from Cleveland, O., where he had been for nearly twelve years previously connected with the Cleveland, Pittsburgh & Wheeling Railroad, nine years of that time as its general freight agent. In March, 1865, Mr. Chandler obtained permission of Wm. Thaw, the originator of the Star Union Line, and its then president, to prepare thirty cars, after a plan of his own, which he believed would prove successful, and, if so, very valuable in the shipment of butter, eggs, cheese, fresh meat, and other perishable commodities.

He called these cars ice houses on wheels, and he is unquestionably the pioneer of the refrigerator car system. Mr. Chandler had not the foresight to patent his idea, which was eagerly seized upon by others." The first of the 30 cars was loaded in Chicago May 16, 1865, with 10 tons of butter, consigned to Porter & Wetmore, New York City, who wrote that the car was a success, and that they were delighted. These first cars were simply box cars, which had put into them an inside lining of boards, which left a space between the new lining and the outside lining of the cars—sides, ends, top and bottom—and this space was filled with sawdust and shavings. At first a box or trough not permanently attached to the car was placed in the middle of the car, and filled with ice. A little later on troughs of this kind, or much like them, were placed in the ends of the car. The Pennsylvania System is justly proud of its position as the pioneer in the transportation of perishable property under refrigeration, which has developed to an extent not even dreamed of by its originator.

In my opening chapter I recorded the impression made upon my ten-year-old mind of the initial use of distinct waste or tankage from the packinghouse of Whittaker & Ryan in Leavenworth, Kans., in the early '70's, and, by the way, years afterwards, when I met C. M. Favorite, one of P. D. Armour's great lieutenants in the Chicago office, he told me that he had worked for Whittaker & Ryan in those early days.

It is a story in itself to give in detail the remarkable part which by-products have played in the packinghouse business up to the present day. I shall deal with them in only a general way. Suppose we begin with bones.

Even in the earliest days with dried beef hams as an important trade factor, bones of the most valuable character went into the general pot of refuse to a great extent. Now they are used for keys on the cheaper grades of organs and pianos, for knife handles, buttons, ornaments and novelties of infinite variety. In order to indicate how thoroughly the whole product is used, I may say that thigh and shin bones are sawed, and the knuckles are used in glue, or crushed and ground for bonemeal, used as a fertilizer, or as a poultry and stock food, and also in the case-hardening of steel. The part between the knuckles is far-reaching. The marrow is extracted and melted into an edible product; the bones are then cooked for neatsfoot oil and tallow, and then dried and sold to manufacturers for the purposes mentioned.

The few fine hairs from the inside of a steer's ear are used to make "camel's hair" brushes. The use of blood is well known, but how many know of the product called "blood albumen"? This is made by allowing blood to coagulate, separating the serum by centrifugal force. The liquid obtained is filtered, decolorized, and evaporated into a dried serum. The final product is used as a clarifying agent in the manufacture of photographic paper, and to set colors in textile printing. During the war blood albumen was used on the wings of aeroplanes to strengthen them, and to protect them from the elements.

I have only used a few striking illustrations. To follow the subject to its finality would be to call in all the trades, all the sciences, and many of the arts into a bewildering labyrinth of what the packer's art, for it is an art, has done in the way of not only saving waste to the world, cheapening the edible product to the consumer, but furnishing to hundreds of other industries materials which have made many other articles possible, and always of such vital use as to cheapen them to the user or consumer.

CHAPTER XXII

ADVERTISING; PARTING REMARKS

DURING my old days in the packing industry I often had occasion to write advertising pamphlets, with some rather odd results. I recall that when preparing an insert folder descriptive of what were then in their infancy "fancy hams and breakfast bacon," such as all packers put out now, and advertise freely in newspapers and magazines, I was stumped for a headline quotation or strong catchphrase. In many instances during my publicity work the same thing had occurred, and when I had exhausted all resources in search of an apt quotation I wrote something myself, crediting it to some dead author or great man, who could not come back to denounce me. I knew that a thing credited to some notable would have more force than if used as original, and I felt that St. Peter would not stop me at the Gate for it, and, if I did get by, I could square myself with the harp-player to whom it had been credited and perhaps make him believe that he really did write or say it. I know it was a low-down trick, but glance back over your own life, brother, and "cast the first stone" if you have never done anything so bad. I heard Henry Ward Beecher lecture several times, and I knew that he could not keep tab on everything that he said, nor could any of the quotation sharks, so I picked Henry, and for my headline wrote:

There is no higher art than that which tends towards the improvement of human food.—HENRY WARD BEECHER.

During the last few years, with spiritism and occult stuff coming back strong among the uninformed, I have attended several seances, and had a "catch-as-catch-can" with ouija, impelled by a guilty conscience for this sort of "clep-to-quota" of mine; though none of those whose names have been burdened with my liberties has ever peeped. But I made Beecher famous in the food advertising line. Jevne of Chicago and Los Angeles lifted my poor little effort, and made it a business headline, but that was not the climax. When we were married my wife and I chose Yellowstone Park for our bridal trip, going via Omaha, and to start things off right I took her into the "swellest" restaurant for dinner. She always sits facing the door, to watch humanity as it pours in, bonnets and all. My face was towards the rear, and I looked twice before I was sure, but there, emblazoned on the back wall, in free relief above the orchestra stand, in letters of gold, was—

There is no higher art than that which tends towards the improvement of human food.—HENRY WARD BEECHER.

I called my wife's attention to the lines, and asked, "Do you know who wrote that?" She replied, "Why of course; don't you see Henry Ward Beecher under it?" Then in my most modest way I explained how it came about, but there was a funny little look in her eyes, and I suppose that together we will have to meet Henry to get the matter straightened out.

Another time I was getting up a catalog for a sale of Herefords to be held by Kirk B. Armour, James A. Funkhouser and John Sparks in Kansas City, Mr. Sparks bringing his cattle from Nevada. It will be recalled that Mr. Sparks bought his basic stock in Missouri, and started a registered herd at Reno, Nev. The herd was in a way forgotten for a long time, until stories of Herefords all over California, Nevada, Oregon and Washington began to come in. There is no doubt that John Sparks' work was of major importance to improved breeding all over the extreme west and northwest. I wanted to convey the idea that the Missouri cattle had gone forth into the wilderness, and after many years returned to the land of their fathers. I felt sure that I would find what I wanted in "Kings," because I especially wanted to credit it to some book in the bible by chapter and verse. I searched in vain. I would not, of course, credit anything I wrote to the bible, because there was some shame left in me, but I conceived a brilliant idea: it was to try to produce something which would have the swing of Kings and express my thought, and to put quotation marks about it, without crediting it to anyone. I worked several days on it, finally getting something which sounded all right, and then I had another brilliant idea: I would try it on some bible student, and see if it would go by. I chose Mrs. Hastings' mother, a clever bible student, and one of the sweetest, dearest women whom I have ever known. I loved her next to my own mother, and felt a little guilty about trying my product on her, but she was a good sport, so I said, "I cannot find this in the bible; can you tell me what book it is from?" Without any hesitation she said, "It is from Kings; I will find it for you." I let it go at that, but she hunted Kings and the concordance diligently before I 'fessed up to the whole plot. John Reid, in charge of our Foreign Department, was also a bible student, so I tried it out on him, being careful to begin with, "I cannot find this in the bible." Offhand he replied, "It is from Kings." The next day he told me that he had spent several hours trying to find it, and again I had to 'fess up.

I have never consciously been guilty of plagiarism and in writing I am careful in the use of quotation marks, but it is strange how quotations, particularly long ones, in the spoken word and properly credited before being used, are accepted so often as original. When the old National Live Stock Association held a meeting in Kansas City about twenty-five years ago its programme committee asked the packers to select someone to reply to the toast, "The Packers," at a wonderful banquet given at the Midland Hotel. I was the victim. That wonderfully gifted lawyer Gardner Lathrop was to preside. The list of speakers got me into a good deal faster company than I needed, so I brushed up as best I could, and in linking up the packing and range ends used Senator Ingalls' beautiful prose poem on "Grass," distinctly crediting him with its authorship. The next day one of my friends said to me, "I liked your address, particularly your wonderful tribute to grass." Here comes your chance, dear reader, for a close decision. Did I tell him whose it was or did I figure that he ought to have known better—certainly should have listened more closely—and let it go at that? There are no prizes in this guessing contest. Suppose you hold the glass up to human vanity, and tell what you would have done.

With the Holiday Number of *The Breeder's Gazette,* my wanderings in memory's world will come to an end in a little story which would be marred by my adding personal comment. I cannot go back to private life, however, without thanking *The Breeder's Gazette's* readers for their generous attention, as evidenced by a mass of letters which I have received from unexpected sources, as well as from old friends, whom I have not heard from for years. One of these letters

—it is from Henry Bonner of Indiana—establishes my truthfulness. He writes: "I have just read your story about old Curlew, the horse that killed Johnny McDuff. I remember the day he pitched with you when you tried to tie the white handkerchief round your neck, when I was looking at cattle on the Tongue River Ranch." It seems odd that as I write of his letter I am sitting in the shade of a Ford car on Tongue River Ranch, near where Curlew pitched, waiting for a bunch of cattle that the boys are bringing up to the shipping pens. In fact, most of these stories have been written on trains, or while waiting about the ranches. It has been a great privilege to write them for *The Breeder's Gazette,* because the backward look has found many things not written down which I have chatted about with men who have long been with us, and, in turn, their own minds take the backward vision, recalling some stirring things that we have been through together. I have a letter from F. D. Coburn, "Coburn of Kansas," which I am putting away in my treasure chest. It is too generous to quote from, but if my little effort had done no more than bring me Mr. Coburn's letter I should feel that I had been richly repaid. I am deeply grateful to the hundreds of others who by spoken word or letter have indulged in kindly comment. There has always been for me the sweetest sort of affection in James Whitcomb Riley's words:

Good-bye, Jim; take keer o' yourself.

In waving my hand from the rear platform of the train which is leaving my good friends in *The Breeder's Gazette* "family," and taking me back to the cattle, the horses, the boys, the birds, the flowers and the silent vastness of the great pastures, I can think of no sweeter thought than in paraphrasing Riley—"Good-bye, reader; take keer o' yourself."

CHAPTER XXIII

THE LOST CHILD

EVERYONE who has lived on the frontier has probably taken part in hunting for lost children. In my own experience and observation the terror, the panic, the tireless zeal, the whole-hearted participation by whole communities are vivid recollections, and as a part of them the never-ceasing wonder at the almost impossible distances children will stray. The natural tendency of an exhausted child to lie down and fall asleep, together with the usual discovery about dusk that the child is missing, and the natural profoundness of a child's sleep, explain to some extent the difficulty of locating it in vast areas, with a belated start, and no trail in the darkness to indicate even the direction that it has taken. Shouting by hunters until they are too hoarse to shout is always a part of the hunt; dogs are always used, and individual stories of their work are traditions over the whole frontier.

It is one thing for a child to stray in a city, where its sobs attract quick attention; it is an entirely different thing when one wanders off into a great expanse, where there are no houses for miles. I have never heard of a child in this country being

Photo by Reeves

Curley and Jack — Story, "The Lost Child"

harmed by a coyote or bitten by a rattler when lost; but those menaces, imagined or real, are the first to come into the mind when a child has strayed, just as a woman grabs her skirts or climbs upon a chair with the first suggestion of a mouse. The people and the incidents in this story are real, but in drawing the sketch I have made it composite as to the lost child terror in sparsely-settled districts, and no amount of dressing can exaggerate instances which have come to my personal notice; in fact, I have toned down the dramatic tragicality which is a part of every instance. The telephone, prevalent now in rural districts, has helped the situation, but it is a terror to operators who after the first alarm reply to every ring, "No, he has not been found;" or, "Yes, they have found him." It was strange how, before the rural system was installed, that which all people on the frontier know as the "grapevine" phone carried the news of "lost child" into great distances. This method of communication, which no one has been able to explain, carried news with startling rapidity.

Little "Curley" McNutt is still a character with us, now seven years old. His imagination is vivid. His laugh and grin, and his habit of saying unexpected things at inopportune times are all still in vogue. I saw him last Sunday, and we had our usual big talk. He was much excited over an air rifle; he has two wonderful "wolf" dogs (really harmless Collies), and their combined kill of imaginary coyotes promises to be big the coming winter. He told me some fantastic tales, and at the close added, "Every time one of us tells a big one, Mac (the name by which he calls his father) tells one a whole lot bigger. He sure is a windy." So perhaps Curley is a victim of heredity. It has just struck me as amusing that local color is not wanting as I scribble, because I am writing in the caboose of a stock train moving through a big pasture country. Just now, while we stopped for water, the "ki-yi" of the coyote's discord sounded off in the mesquite. I wonder whether the cry is striking terror to the heart of some mother whose baby may have strayed. There are no hoarse shouts, no ghostly riders in the pale moonlight, and that strange Providence that finds the lost ones when they stray is watching over the tired little boys and girls, fast asleep, straying perhaps with Peter Pan in his frolics.

Curley McNutt was at the time of this incident four years old, a round-faced, flaxen-haired, sturdy little fellow, wearing a perpetual grin. He had a habit of repeating what he heard one say, prefacing it by a short laugh. All ranch children make warm friendships, which often amount to idolatry among cowboys, who in turn have an absorbing love for children. This love sometimes drifts into a sort of slavery.

The "doghouse," a classic title which the boys have bestowed on their home, the bunkhouse, is a point of absorbing interest for all ranch children. Here they are teased and petted and properly spoiled; no fence, latched door or set of material regulations and no amount of punishment are effective in keeping children away from the "doghouse;" after they begin to toddle, and commonly, long before that period they are carried there by the boys to learn the delights, the privileges, and their own power of tyranny.

I have said some pleasant things about cowboys in this series, but candor and a desire to be truthful force me to admit that often, in the moral and intellectual environment of the "doghouse," a child may pick up some chance word or expression which is not strictly good form. I have known of that possibility, however, in more effete circles—afternoon teas and ladies' bridge clubs, for instance, and even Sunday school. But I have never known a cowboy deliberately to teach a child to "cuss." Any one who has had much to do with handling cattle knows that at times there is no other outlet than a little mild profanity. I have even fallen into it myself, in distressing circumstances, and once I heard a preacher swear.

Curley often had his mouth washed with good clean home-made soap for bringing certain gems into the family circle after prolonged visits to the "doghouse." He had a habit, too, of repeating some inoffensive phrases which, falling in the wrong place, were sometimes disastrous. I recall an incident which was only saved by the common-sense, the infinite good nature and sense of humor of the victim. When Curley was allowed to help himself at the ranch mess table one of the boys would say, "Don't take it all"—a comment which Curley soon learned to make when cowboys took a helping. It was sometimes a wise admonition, as any one who has lived around a cowcamp knows.

I took a buyer of goodly quantities and his wife to the Tongue River Ranch. Both were charming. They had brought their outdoor appetites with them. Mrs. McNutt is one of the best cooks I have ever known, on or off the ranch, and the supper that night was one of her best triumphs. We had ridden 100 miles that day over rough country with Mr. Ford, and were ready. The lady had "thrown in" with Curley before supper, and he had to sit by her. The first helping "got by," but when the lady came up for a second one Curley was primed, and out of the stillness which prevails generally at an indoor cowboy mess table came his high, clear voice with "Don't take it all." For a second the spoon poised in mid-air; the cowboys, appalled by their tutorship, were tense; then followed the merry, ringing laugh of a splendid woman, as the spoon descended to its mission and she said, "All right, Curley; but you and I must get our share before the cowboys 'take it all'."

Curley's father and mother had charge of the S. M. S. headquarters messhouse on the Tongue River Ranch. The great part of "Mac's" time was taken up with near-by line or pasture riding, leaving the mother and boy often alone for much of the day with the cow outfit off on its work. Curley always thought of the "doghouse" as the objective when he could make his "getaway"; but when he could find no one there he had a habit of striking out straight into the pasture, in all its uninhabited immensity, and several times had wandered some distance, giving his mother much anxiety; but she felt perfectly easy when any of the boys were about. "Old Jack," a cross between wolfhound and greyhound, has gone to his fathers; but he is still a tradition on the S. M. S. Ranch. He followed the Tongue River chuck wagon for ten years, with some memorable coyote encounters. He was often away with the wagon for weeks; sometimes he would let it go away without him, and then unerringly locate it in some remote part of the pasture or sometimes he would leave it and come to headquarters, evidently to see Curley. They were great "pals," and so close was the companionship between them that it was hard to tell which had the more fleas. They ate out of the same plate, when the chuck wagon was at headquarters, and Jack always played fair.

Late one afternoon Mrs. McNutt saw some cowboys ride in, and supposed that they were at the bunkhouse. Curley disappeared soon after. She saw him making for the bunkhouse, and thought he was safe for the rest of the day, but did not know then the boys had passed through and gone on some special mission. It was in May. Copious rains had fallen all spring; the growth of vegetation was prolific everywhere, almost rank. No one who has not seen a west Texas landscape, after a wet spring, can form any idea of how the growth would obscure a child walking in it. While it is not a part of my story, I should like to add that no one who has not seen it can conceive of the wonderful floral beauty of the country. There are vast expanses of chrome-yellow, clear to the horizon, variegated here and there by acres of soft lavender, or brilliant purple, looking in the great vista like a single vivid flower of contrast pinned to a woman's dress. As one studies more intimately the carpet of color at one's feet, one sees a riot of white and yellow primroses and daisies; the deep rich wine-tint of wild hollyhock; the blue of larkspur and star daisy; and the magenta of wild verbena. One is lost in the wonder of nature's flower show, held way off in the wilds, with God's love for its reward. I have never been entirely in sympathy with the beautiful lines—

Full many a flower is born to blush unseen,
And waste its sweetness on the desert air.

I watch the grass birds as they nest in the profusion of beauty, and the bees as they gather honey. The perfume of flowers and grass, as vagrant breezes drift it here and there, must be a joy to every living thing.

Supper on the ranch usually comes at dusk. The bell rang, and the boys in their silent way drifted into the messroom, but there was no childish voice. The mother appeared in the kitchen door with the query, "Where is Curley?" No one knew; no one had seen him. Then a face went white, followed by solemn-visaged cowboy faces that through their bronze grew even whiter. Supper was forgotten. Every man dashed for his night horse. The discipline of the ranch came quickly; a silent mounted group drew round the strawboss; there were quick, clear orders as to districts and signals. Then "Molly" O'Hare, the horse wrangler, Curley's slave, said, "Call Jack; he can find him." Some one called, but Jack did not answer. Then there was a shout of joy, "Jack is with him; he is safe."

As the band rode off there came the first sharp, shrill "ki-yi" of the coyotes. One makes enough noise for a pack. To the lost child hunters there seemed to be thousands of coyotes. From every direction they seemed closing in. It was as if thousands were sounding their hungry call on the night air. Was Jack with Curley? Jack was getting old. How long could he fight off the coyotes? Would one get Curley while Jack fought the others? Would Curley step on a rattler and be bitten?

All the frenzied fancy of human love alarmed marched in hideous phantasy before them as they searched. From hill and dale came the peculiar rounding cry of the cowboy, but there was no signal. They rode carefully and closely covering the ground. Hushed voices answered hushed voices as they crossed on the hunt. The night was clear, but dark; the moon was not yet up. The coyote chorus grew louder; they seemed to be gathering from afar, and "Molly" O'Hare, riding alone, cried in the agony of his heart, "O God, if I only knowed how to pray I'd tell you to take me and let Curley stay; I'd tell you that there ain't nothing you could tell me to do that would be too much; I'd—"

Then out of the stillness of the night came a different note; a note that said, "I am here;" a note that said, "All is well;" a note that said, "I do not fear." It was the long-drawn-out bay of the wolfhound's answering call. It halted a dozen riders; they waited to hear it once more, and it came clear and sure, "All is well; we are here." It jarred the stillness with its wave of triumph.

The moon came slowly up as a frenzied bunch of riders on horses which knew something was wrong broke into a killing run. "Molly" O'Hare was the first man in; Curley lay stretched out on a mat of wild flowers, fast asleep. His little shoes were worn through, his blouse and knickers ragged from contact with catsclay bushes, his stockings hanging down, showing angry scratches on his little legs. Beside him, alert, head up, his muzzle to the wind, still sounding his answering note, stood Old Jack.

With a bound "Molly" O'Hare was by their side. One quick caress for Jack, and Curley was in his arms; then, with characteristic cowboy nonchalance, he turned to the bunch of riders, leaning forward on their horses' necks in a circle about him, silent and indifferent as though nothing had happened, and said, "Oh, hell! I knowed he was all right all the time." Whereupon, Curley, half aroused for a moment and from long force of habit, repeated, "Oh, hell, I knowed he was all right all the time," and fell into another sound sleep.

The boys with great glee told his mother of Curley's repeating "Molly's" exclamation, but holding him close to her breast, tears stealing down her cheeks, and dropping on Curley's torn raiment, she said, "Poor tired baby, he was too sleepy to know what he said; I guess we won't wash his mouth out with soap this time."

CHAPTER XXIV

"OLD GRAN'PA": A "MAGE" STORY

IT WAS early in June, 1906. The "High Boss" was out from New York, making his usual semi-annual inspection of the S. M. S. Ranches. We had already been over several of them, and were getting into the buggies to drive from Stamford to Throckmorton when *The Breeder's Gazette* man showed up, with only one day to give us. Those were the days of Hynes Ranch buggies, and "Broom Tails." Throckmorton Ranch was a full day's drive away; now Henry Ford takes us there in 2½ hours. Flat Top Mountain Ranch begins 5 miles west of Stamford. There was my only chance to show *The Breeder's Gazette* man anything, so I told our party to go without me, and I would drive through that night in ample time to start on the Throckmorton inspection trip next morning. On the way over to Flat Top Ranch I stopped at headquarters, and asked Mage the foreman to meet me in town at 7 p. m., and we would drive Beauty and Black Dolly, two spanking mares which he had bought for me. They could take their 10 miles an hour steadily for hours, and I threw them in as a bait to tempt Mage against any local duty which he might urge. Mage stood 6 feet 5

Photo by the late Eric A. Swenson

Mage himself on "Old Gran'pa"

inches in his socks, every inch of it cowman and horseman. He came to the ranches at thirteen years of age—a much misunderstood kid. But he had grown into a manhood of sweetness and strength, which had surrounded him with the love and respect of every man, woman, and child in the country. Mage was a dead-game sport, a rider whose skill and daring are still traditions in the big-pasture country. His stories and personal reminiscences, told with rare humor and dramatic force, made a journey with him a real entertainment. I always sparred for an opening to get him going when we made drives together. At 7 o'clock he was on hand to the minute, talking to the mares as though they were human. We were off—"heads up and tails over the dashboard." As we swung into the main thoroughfare the people on the street turned round to watch Mage handle the mares. They were having their little fun before settling to the steady distance-killing gait, and they were a pair to look at: Beauty a deep chestnut, both wilful and beautiful, and Black Dolly, with her sleek sable coat, still at the giddy age. Mage had the stage driver's trick of coming into town or going out in style. The mares knew his voice and hand, and the light that shone in his eyes told where his heart was. For two hours we chatted or were silent by spells, as is the habit on long drives. The moon came up in her soft fullness—one of those southern moons like the ripeness of love, a perfect heart full.

The cool night air was stirring caressingly, and we were both under the spell of it all. The mares had steadied down to normal. We were crossing a prairie near Rice Springs, once a famous roundup ground in the open range days. Mage raised his six-feet-five up in the buggy, looked all around, and, as he sat down, said: "This here's the place; here's where me an' Old Gran'pa won our first ditty."

The moon had risen high enough to flood a great flat until we could see a mile or more. I saw just a beautiful expanse of curly mesquite grass, blending its vivid green with the soft silver moonlight, but Mage saw great crowds lined on either side of a straight half-mile track; two riders; the one on a midnight black and the other on a speed-mad sorrel, in deadly contest for supremacy. The stillness of the night—which to me was the calm benediction of peace and rest—was broken for him by wild cheers as a boy and a sorrel horse crossed the line, victors. His face was tense, his eyes shone with the fire of strain and excitement, and then slowly he came back to the stillness and to the moonlight, and to me.

I waited a minute, and asked, "What was it Mage?" He did not answer until we had crossed the flat. Then, with a little short laugh, peculiar to him before telling a story, he began: "As fur as thet's consarn it wus this away—"

But here let me tell some true things I knew about Old Gran'pa. He was a famous cow pony, originally known as Sorrel Stud. Mage broke him as a three-year-old, and had ridden him some eighteen years. The last few years of that time Stud had come to be known as Old Gran'pa. He was still alive, but had been turned out under good keep, winter and summer, to end his days in peace. He was very fast, and was considered among the top cutting horses of his time. Mage's worship of this horse is only typical of every cowboy's love for his pet horse. But to his story:

"It wus this away: We hed fenced some, but allus hed lots o' strays on the open range, an' Shorty Owen [who, by the way, stood 6 feet and 6 inches], tole me early in the spring he would send me out to gather strays when the big round-ups begin, an' 'lowed I best be gettin' my plunder rounded up. That wus 'fore you cum, but you know he wus the S. M. S. range boss, an' mighty nigh raised me. He tuk to me the day I hit the ranch. 'Kid,' he says, 'you ain't never hed no chanct an' I'm agoin' to giv you one.'

"Shorty taught me to ride—hobbled my feet unter a three-year-ole steer onct, an' turned him a-loose. We hed it roun' an roun' with the hole outfit hollerin', 'Stay with 'im, Kid!' I stayed all right, but when he pitched into a bunch o' mesquites I sure would 'a' left 'im if these here preachers is right 'bout 'free moral agency,' but them hobbles helt me back, and I stayed fer the benediction. Since thet time I never hev seed a hoss I wus scart to climb on.

"Shorty cut Sorrel Stud out to me when he wus a bronc, an' said, 'Break him right, Kid; I think you got a cow hoss if he ain't spoilt in the breakin.' An' I done it without ever hittin' him a lick. As fur as thet's consarn, I never did hit him but onct, an' thet wus the time him an' me both failed, only Shorty said we didn't fail; we jes' went to the las' ditch. But thet's another story.

"I wisht you could a-seed Sorrel Stud in his prime. He wus a hoss! I thought 'bout it today when you hed yore arms round his neck an' a-talkin' to him 'bout me, an' I wondered if any body 'cept me could understan' thet Sorrel Stud and Ole Gran'pa wus the same hoss. But when I got up an' thumbed him, an' made him pitch me off jest to show you what a twenty-year-ole hoss could do, did you see the fire come into them eyes, an' them ears lay back? Hones' to God, Frank, he wus a hoss!

"I know I wus jest a tough kid when I come, but a-tween Shorty Owen an' maybe a little doin' right fer right's sake I tried to live an' hones' life. But they's two things me and St. Peter may hev to chew 'bout a little at the gate. You know what a fool I am 'bout tomatoes? Well, onct I stole a dozen cans from the chuck wagon and hid 'em out in the cedar brakes. But the boys at the wagon hed me so plum scart 'bout Injuns thet I never did git to them tomatoes. Well, Ole Gran'pa is jest as plum a fool 'bout oats as I be 'bout tomatoes. I'll admit I stole this here outfit's oats fer him ten years, till the High Boss wus out onct from New York and seed Ole Gran'pa go to a fire. Of course I wus up, an' he sed he guessed he could pay fer Gran'pa's oats the rest o' his days. Joe wus mighty perticular 'bout company oats. We hed to haul 'em 60 miles, but I think he slipped a mess to White Pet onct in a while hisself. I used to wait 'til the boys hed hit their hot rolls, then I'd slip out to the barn, get my big John B. full o' oats, steal to the corner o' the hoss pasture, an' Ole Gran'pa wus allus waitin' fur me an' he'd never leave a stray oat to give us away.

"They called me 'the S. M. S. Kid.' I wus 'bout sixteen. I could ride some an' I allus hed a little money back from my wages. So when Shorty Owen tole me I wus agoin' I used thet an' all I made up to goin' time fer an outfit. I hed a good season saddle, a Gallup; but I bought a bridle with plenty o' do-dads on it. Then you know my Injun likin' fer color: I bought a yaller swet blanket, an' a top red Navajo blanket fer Gran'pa. He kinda leaned to color too. I set up all night with Swartz an' made him finish a pair o' top stitched boots, an' I hed enuff left fer new duckin' pants, red flannel shirt, an' a plaid fer change, shop-made bit and spurs, both inlaid, a yaller silk handkerchief, a new hot roll, an' a twelve-doller beaver John B. Then Shorty Owen cut out my mount. In course I hed Sorrel Stud; he wus six years old, right in his prime, an' I kep' him shinin'.

Then there war nine more, all good ones—Blutcher, Alma, Polecat, Tatterslip, Bead Eye, Louscage, Possum, Silver Dollar, an' Badger, three of 'em from Shorty's own mount.

" 'Kid,' says Shorty, 'you got as good as the best o' 'em. I wants fer you to mind thet on this here work you're representin' this here outfit. Keep yore head, an' come back with it up. But I'd bet my life on you, an' this here outfit is trailin' you to the las' ditch'."

Mage's voice was getting low here, and he swallowed on the last words, paused for a moment, then with that laugh of his continued: "Well, I'm stringin' 'em out a mile here, when I ought to have 'em bunched. Thet wus a great summer. I worked in the big outfit with men an' hosses thet knowed how to turn a cow, an' the captain o' the round-up got to puttin' me an' Stud into the thick o' it purty reg'ler. It allus seemed thet when I rode Stud, Split Miller rode a little hoss called Midnight, an' he sure wus a hoss; black as midnight, 'cept fer a white star in the forehead, short-coupled an' quicker then forked lightnin'. He would cut with the bridle off, and fast? He was a cyclone. Every night 'roun' the camp fire Split kep' pickin' a load in to me 'bout the Stud. Onct it wus, 'Well, Kid, I seed you hed the little scrub out watchin' Midnight work.' Or, 'Say, Kid, I believe if you hed somethin' to ride you'd be a hand.' I swelled up some, but I 'membered what Shorty Owen sed, 'Keep yore head an' come back with it up.' An' Split wusn't mean. He jest luved to josh. Two or three times the captain said, 'Split, let the Kid alone.' But he'd shoot one at me as he rode by in the work, and wus allus badgerin' me fer a race.

"Then I kinda fell into watchin' Midnight run somethin'; an' I'd start Stud in the same direction to pace him. An' I cum alive; the Stud was full as fast. I jest naturally supposed thet Midnight could beat anything, but I kep' a-tryin' an' my eyes kep' a-openin'. One night Split got mighty raw, an' finally says, 'Kid, I'll jest give you twenty dollers to run a half-mile race, standin' start, saddle agin saddle.' An' then I fergot Shorty's instructions an' los' my head.

" 'Split,' I ses, 'you been pickin' on me ever sinct I cum to this here work. Me an' Stud don't need no twenty dollars to run you. An' even break's good enuff fer us, saddle fer saddle, bridle fer bridle, blanket fer blanket, spur fer spur.'

" 'Good enuf, Kid,' ses Split, 'got anything else—eny money?'

" 'No,' I ses, 'I ain't got no money, but I got sum damned good rags an' a new hot roll.'

"Then the captain o' the roundup tuk a hand. But my blood was up, an' they put cash allowance on all my plunder an' I bet it 'gainst money. They give me $12 fer my Swartz boots, $8 fer my John B., $5 fer my cordaroy coat, $4 fer my shirts, an' $2 fer my duckin's. It war Wednesday, an' the race wus to be pulled off Saturday evenin', straight half-mile, standin' start at the pop o' a gun. The captain tuk the thing in charge an' sed he'd lick eny damned puncher thet tried to run a sandy on the Kid. It was all settled, but by the time I hed crawled into my hot roll thet night I 'membered the talk Shorty Owen give me. Stud wus kinda mine, but he war a company hoss, arter all, to work on an' not fer racin', an' I sure wus in a jackpot fer losin' my head. Well, the nex day I tuk Stud off to practice fer a standin' start. You know how I say 'Now!' when I'm workin' on a hoss and jest as I want him to do somethin'. Well, Stud he'd been trained thet a-way, with jest a little touch o' the spur, an' I figured to say 'Now!' as the gun popped an' touch him thet a-way, an' he got the idee.

"Thet night I tuk him to the track an' put him over it four or five times. An' onct when we wus restin' a-tween heats I says to him, 'Stud, if me an' you loses this here race looks like we'd hev to steal off home in the night an' both o' us mighty nigh naked.' Everybody knocked off work Saturday. You know how even in them days word gits 'bout by the grapevine. Well, by noon they wus ridin' and drivin' in from all directions. The wimin folks brought pies and cakes. The cusey cooked up two sacks o' flour an' we hed to kill two beeves. Everybody et at the chuck wagon an' it wus sum picnic. I tol' the fellers not to bet on me an' Stud, but they wus plenty o' money on both sides. An' a girl with black eyes an' hair an' jest as purty as a bran' new red wagon, ses, 'Kid, if you win I'm agoin' to knit you sum hot roll socks.' An' Ole Pop Sellers ses, 'Better look at them feet an' begin figurin' on yarn, 'cause the Kid's agoin' to win.' But Split hed a girl, too, an' she up an' ses, 'If the Kid's dependin' on them there socks to keep warm he's mighty apt to git frost-bit this winter.' Well, you know the josh thet goes 'round when a big bunch o' cow people git together. An' they wus a plenty, until I wus plumb flustrated. When the time cum, a starter on a good hoss wus to see thet we got off fair an' then ride with us as sort o' pace-maker an' try an' see the finish. But his hoss wusn't in Midnight's an' Stud's class.

"Split hed seemed to figure thet Midnight didn't need no trainin', he hed run so meny races an' never been beat. So all Split did wus saddle Midnight and stan' 'round an' josh. But me an' Stud was addled, an' I warmed him up a bit, talkin' to him all the time. I wus worited 'bout urgin' him in a tight place. I hed played with my spurs on him, but he never hed been spurred in his life 'cept a signal touch to turn or jump. I allus carried a quirt on the horn o' my saddle, but 'cept to tap him in a frenly way or in work he hed never knowed its use. What wus I a-goin' to do in a pinch? I knowed he would use his limit under my word, but what if he didn't? Did I hev to hit him? If I owned this here ranch I'd hev give it all to be out o' the race an' not look like a quitter. Well, the time wus cum. Stud hed been frettin' an' I wus stewin', but when we toed the line sumthin' funny happened: We both seemed to settle down an' wus as cam as this here night. I jest hed time to give him one pat an' say, 'God A'mighty, Stud, I'm glad I got you,' when the starter hollered, 'Git ready!' An' the gun popped! I yelled, 'Now!' at the same time, an' we wus off.

"Midnight wus a mite the quickest, but Stud caught his neck in the third jump an' I helt him there. I wanted Midnight to lead, but kep' pushin' him. We didn't change a yard in the fust quarter an' Split yelled, 'Kid, yer holdin' out well, but I got to tell you farewell.' An' he hit Midnight a crack with his quirt. Stud heard it singin' through the air an' jumped like he wus hit hisself. In thirty yards we wus nose an' nose; ten more, a nose ahead. Then I knowed we hed to go fer it. I wus ridin' high over his neck, spurs ready, my quirt helt high, an' I kep' talkin' to him an' saying, 'Good boy, Stud!' The crowd wus a-yellin' like demons. We wus in the last eighth, nose an' nose, an' I let out one o' them Injun yells an', 'Now, Stud! Now!'

"It seemed like he'd been waitin' fer it. I could feel his heart beatin' faster. There wus a quiver wint through him like a man nervin' hisself fer some big shock. An' I could see him gainin'—slow, but gainin'. The crowd hed stoppt yellin'. It cum sudden. They wus so still you could hear 'em breathe. I guess we must a-bin three feet ahead, with a hundred yards to go. Split was a-cussin' an' spurrin', an' whippin'. I didn't hev no mind to yell in all thet stillness, I wus ready to spur, ready to whip, an' my heart wus a-bleedin'. I don't think now thet I could a-done it to win, an' I jest whispered, 'Now, Stud! Now! Now!'

"I thought he wus a-runnin' a-fore, but he shot out like a cry o' joy when a los' child is foun'; an' we crossed the line a length an' a half ahead. I seed the black-eyed girl with her arms 'round Pop Sellers' neck an' a-jumpin' up an' down. Pop wus jumpin' too, like a yearlin', an' the crowd wus doin' an Injun dance generally. Stud didn't seem to sense the race wus over, an' wus still hittin' the breeze. I checked him in slow, pattin' him on the neck, an' talkin' to him like a crazy man, 'til he stood still, all a-quiver, his nostrils red as fire an' eyes still blazin'. Then I clum down an' throwed my arms 'roun' his neck and ses, 'God A'mighty, Stud, I didn't hev to hit you.' Stud's eyes seemed to softin' an' he laid his head down over my shoulder. I wus cryin' like a baby, huggin' him hard. The boys wus ridin' to us an' Stud raised his head an' whinnied. I guess it wus jest the other hosses comin', but I thought he sed, 'Didn't we raise hell with 'em?' An' I ses, 'You bet we did, Stud, but it wus you done it.'

"News travels fast, an' long 'fore I got in with

my strays they knowed all 'bout it at headquarters. I kep' thinkin' 'bout what Shorty sed, 'Come back with yore head up,' but I hed mine down when he met me at the corral. I knowed we hadn't no hosses to race fer money. He looked kinda hard at my extra saddled hoss an' roll o' plunder and ses, 'Kid, this ain't no racin' stable. This here is a cow outfit, an' our best hosses is fer cuttin', not racin.' I didn't say a word, jest unsaddled an' started fer the dog-house, when I herd him cumin'. He caught up with me, grabbed me by both shoulders an' turned me 'roun'. I saw a great big tear stealin' down his cheek, an' he ses, 'God A'mighty, Kid, I wisht you wus my boy.' Then he turned away quick an' wus gone, while I set down on the groun' an' blubbered in my ole fool way thet I hev never got over. When pay day cum Shorty handed me my wage check, which had growed sum, an' sed, 'Kid, when a boy does a man's work he gits a man's pay. You begin doin' a man's work when you went to gather them strays, an' you cum back the same way.'

"Then he started to go on, but turned and sed, 'Say, Kid, if I owned this here S. M. S. Ranch, hosses an' cattle, I'd a-give the whole damned outfit to a-seed you an' Stud cum over thet line'."

CHAPTER XXV

THE SPECKLED YEARLIN'

APRIL and May rains, followed by good growing weather, had made everything beautiful in the S. M. S. pastures. The turf of curly mesquite grass was like a beautiful rug, painted here and there with wild verbena, star daisies, white and yellow primroses, and the myriad coloring of west Texas flora. Branding time was on, and the S. M. S. Flat Top Mountain outfit had gone into camp at Coon Creek Tank, to begin work the next day.

"Scandalous John," the foreman and wagon boss, had been through the aggravating experience of getting an outfit together. It had been no trouble to find riders—cowboys who knew the game from start to finish—but to secure a cook, a "hoss wrangler" and a hoodlum wagon driver was a problem. No one wants to drive the hoodlum wagon, with the duties of supplying wood and water for camp and branding, helping the cook with his dishes or other odd jobs, unprofessional, from a cowboy standpoint, except so far as they lead to a "riding job," meaning regular cowboy work. The "hoss wrangler" was not hard to find, but whoever takes the job aches all the time to be promoted to a riding job, and is therefore dissatisfied. The hoodlum driver had worked one day, and quit. Scandalous was racking his brain to know where to look for another, and was saddling his horse to hunt for one when Four-Six, one of the cowboys, exclaimed, "Look what's comin'!"

Along the dim pasture road, miles from any dwelling, a figure on foot was approaching—a sight which always attracts attention in the big pasture country, since it is associated in the public mind with suspicion, if the footman is unknown. It often occurs that some one's horse will get away or give out. The rider then makes for the nearest cow camp to borrow a horse; but a man walking needs some explanation, although he is always fed without question. The boys were all quiet and indifferent, as they commonly are in a cow camp when a stranger approaches.

A lad of sixteen, rather the worse for wear, clad in a shirt and ducking trousers, badly frayed, a soft felt hat, full of holes, shoes badly run-down at the heels, and bare toes showing through the uppers, stopped within ten feet of the wagon. Scandalous paused in his saddling to say, "Well, son, in trouble?"

The lad's face, lit up by a broad grin, made an appeal to the whole outfit, and all were at attention for his answer. "No, I'm looking for the S. M. S. boss. They told me at the ranch house that he was here, and I'm looking for a job."

"You look hungry, son; come eat, an' then tell us all about it," said Scandalous.

As the lad ate, and refilled his plate and cup, the cook ventured, "Son, you're plumb welcome, but when did you eat last?"

"Night before last," the boy replied. "The brakies give me some bread and meat, but I sure was gittin ready to eat when I smelt your grub cooking down the road."

"Where be you from, son?"

"I'm from Virginia," came the reply, "and I'm sure glad to get here, and get a job."

"Virginia! A job?" exclaimed Scandalous. "How did you get here, an' how do you know you kin get a job?"

Again that good-natured grin appeared as the lad told his story.

"I walked some, and rode with the brakies some; they was mighty good to me, and give me a card to other brakies; sometimes they'd give me food they cooked in the caboose, and sometimes they took me home. I told them I was coming to the big S. M. S. Ranch to work. I worked on farms some, but hurried as much as I could, to be here branding time. Am I in time?"

The quiet assurance of the boy staggered Scandalous, but he recovered to ask, "How did you know about the S. M. S. Ranch? What made you think you could git a job? Ever done any cow work?"

The lad's grin broadened as he answered: "Well, a feller I worked for down in Virginia had one of them picture books about the S. M. S. Ranch, and I read where it said, 'No use to write for a job,' so I just cum. I kin do anything I start out to do; I wanted to work on a ranch ever since I was a little feller; I can learn to do anything you want done, and I sure am going to work for you."

Scandalous blinked again, and said, "Why, son, we would hev' to hev' permission from your pa and ma, even if we had a job, 'cause you might git hurt."

A shade of sadness swept for a moment over the young face; then it shone again with a new light of conviction.

"I ain't got no pa or ma, I been in the orphan asylum until two years ago, when a fine man, the one with the book, took me on his farm to do chores. I didn't run away from him, neither; he said I was so crazy about comin' I'd better start. I been on the road so long the things he give me wore out. I guess I walked about a month. They told me in town to go to the office, but I was afraid they'd turn me down, so I cum to camp, and I'm a-going to stay and work for nothing."

There is a straight path to the hearts of cowboys, if one knows the way, and Scandalous was glad to hear the chorus from the whole outfit, "Let him stay Scandalous. We'll help him. Give the little boy a job."

"Reckon you kin drive the hoodlum wagon, 'Little Boy'," said John, and, like a flash, came this response: "I don't know what a hoodlum wagon is, but I kin drive it."

It was settled. "Little Boy" was hired, and "made good." Every moment that he could get from his work found him in the branding pen, and, as is the custom with cowboys in their work, he often rode big calves. The boys, watching his skill, would get him to pull off "stunts" for visiting cowmen, until it began to be noised about that "Little Boy" in the S. M. S. outfit "was sum calf-rider." Then came the proud day of his life, when an older man was found for the hoodlum wagon. The horse wrangler was promoted to a riding job, and "Little Boy" to horse wrangler.

The boys had from the outset contributed shirts and socks; ducking trousers had been cut off for a makeshift. The first month's wages had provided a fair outfit, including the much-coveted white shirts that cowboys love to have in their "war bags" for special occasions. Succeeding months brought saddle, bridle, spurs, horse blanket and a "hot roll." "Little Boy" was coming on, but had to content himself with shoes until he had all the major necessities, and could acquire the two grand luxuries: a \$15 John B. hat and \$35 hand-made stitched top boots.

All through the summer "Little Boy" progressed, first from calves to yearlings in his play time, and then to outlaw broncs, until the boys in the outfit would say, "Thet kid sure kin ride; I'll bet he gets inside the money this fall at the Stamford rodeo."

Anything pertaining to an outlaw horse or steer becomes current gossip in the big pasture country, where horses and cattle form the basis of conversation about the wagon after working hours. Strange stories drifted in about a certain outlaw speckled yearling on the Lazy 7 Ranch—he had thrown every boy with rodeo aspirations who had tried to ride him, and seemed to be getting better all the time. The "Speckled Yearling" was tall, gaunt and quick as a cat. He had a mixed jump and weave that got his men about the third jump, but the boys on the Lazy 7 were keeping him to themselves, with a view to pulling off a prize "stunt" at the Stamford rodeo in September. All the little country towns held rodeos during the summer, with calf and goat roping, bronc-busting and steer riding, but the big event was to come, and the boys were getting ready for it. "Little Boy" had a heart-to-heart talk with his boss, and received permission to ride steers, and tackle the "Speckled Yearlin'," if opportunity permitted.

At last the time for the great event came. Cowboys from 100 miles around were on hand. Professionals were barred. It was to be an event for boys who were in actual service on ranches. The S. M. S. headquarters office was thrown open for all, and the Stamford Inn pulled off an old-time cowboy dance, with old-fashioned "squares" called by old-time punchers, with old-time fiddlers doing the music. The weatherman had done his best; some 2,000 people filled the grandstand, cheering the events of the first day, with now and then a call for the "Speckled Yearlin'," which was not mentioned in the programme.

Any one who has not seen an unprofessional rodeo knows little of real cowboy sport, since it differs in its wild abandon, grace and skill from the staged events. As each favored son came on for his "stunt," he was cheered to the echo, and usually he pulled some original antic which sent the crowd wild.

The announcer, riding before the grandstand, waved for silence. "Listen, people: I want you to hear this; it's a surprise, and the big event. No one has ever been able to stay ten jumps on the 'Speckled Yearlin',' from the Lazy 7 Ranch. Nig Clary will now 'ride at' the Speckled Yearlin' on his own risk: A $50 prize if he stays on; a $25 forfeit if he gets throwed. If he rides him down, a hat collection will be took. If Nig can't ride him some other feller gets a chance tomorrow."

"If Nig can't nobody kin," shouted the grandstand. "Turn him a-loose." A wave from the judges' hands, and, like the cutting off of an electric current, all was still and tense. Then from the mounting chute shot the "Speckled Yearlin'," with Nig Clary up, clinging by two hand-holds to a surcingle and riding bare-back. The yearling was dead-red, with distinct white speckles about the size of one's thumb distributed well over his body. He carried long, sharp horns; his back was on the order of an Arkansas razorback hog. When it came to jumping and weaving his body at the same time, the "Speckled Yearlin' " was the limit.

Nig sat straight for three jumps, began to wabble in the fourth, and was on the ground at the fifth. Still jumping, the yearlin' turned and made for him, giving Nig only time by a scratch to climb up behind one of the judges.

The second day found "Little Boy" and Scandalous with their heads together. "I know I kin ride him, John, an' I sure want that prize money for my boots an' my John B. They's all I'm needin' to be a real cowboy."

"Yes, I know," said John, "but we're needin' live cowboys, an' I ain't feelin' right 'bout your tryin' thet yearlin'. I'll hev to ask you to waive all blame fer the company, an' if you do git hurt they'll be blamin' me; but if you be bound to ride, us boys will pay the forfeit, if you get throwed."

Again on the second day the announcer waved his hand for silence. "Folks, yesterday the best rider and cowpuncher in Texas rode at the speckled yearlin'. Today 'Little Boy' from Flat Top Mountain Ranch says he's goin' to ride him. We hates to let a little orphan boy go agin this here steer, but he sez he ain't a-goin' to git hurt, an' if he does there ain't anybody but him. The management hopes he wins.

If he does, git your change ready for a hat prize, an' I am a-goin' to start it with a five."

As boy and steer came out of the chute, the stillness fairly hurt. Every heart in that great crowd seemed to stop for the first three jumps, but "Little Boy" was sitting tight. From the crowd there came a mighty roar: "Stay with him, 'Little Boy'! He's got a booger on him. Ride Him 'Little Boy'!"

At the tenth jump "Little Boy" was still up, his grin growing broader and his seat getting steadier, while the yearling, maddened by his clinging burden, pitched and weaved, but, like Sinbad's "Old Man of the Sea," "Little Boy" kept "a-ridin'."

The crowd went daft. Every one was standing and shouting. The noise seemed to infuriate the yearling, and, turning from the end of the enclosure, he made straight for the grandstand, struck his head against the protecting wire, stood stock still, and glared, while "Little Boy" sat and grinned. Some one cried "Speech!" and, as stillness came, "Little Boy," still sitting on the dazed steer, broadened his grin and said, "I jest had to ride him. I needed them boots and thet John B., so's I could be a real cowboy, an' this yere speckled yearlin's done done it."